以奢为美
新古典豪宅设计

中华文化 策划
陶陶 主编

NEO-CLASSICAL VILLAS

江苏人民出版社

Preface I
I Have to Say

Recently I write some prefaces or forewords for a number of books or magazines almost every month, such written works make me have more concerns, because the prefaces and forewords are the overall summary of books, it's quite easy to mislead readers if the work done is not quite appropriate. I also know that I myself is very emotional, and I followed my true inner feelings to write essays. Ever since I began to write essays, I had never disobeyed my conscience and social morality, nor write for money.

I can't remember how many essays I've written since I began seven years ago. Nowadays the editorial standards are higher and higher, and the readers are having more sophisticated taste, so it's hard to please them if the work is not done by following true feelings. Meanwhile those interior design publications are uneven in quality, thus it's more and more difficult to find a really good book, and I'm almost out of good words to say when I was invited to write some prefaces and forewords. I still remember this year I wrote a preface without high recommendation with some of my personal opinions on the quality of the book layout and design. It's unbelievable that there's such brave editor that put every word I wrote published. I really admire the courage, and I have to admit I underestimated the editors.

People tend to put some good words when it comes to forewords and prefaces, and most of the books about design have similar forewords and prefaces, thus which are ignored, and the readers will go through the content and make their own judgement to determine the quality of the work, and then decide whether to buy it or not. Ever since I started writing forewords five years ago, some publisher friends told me that I already had some design fans for writing essays, and also told me there are many designers who look at the essays I wrote and then decide whether to buy the book. Amitabha, I have to say, don't blame me if the book is not the best seller because of the forewords I wrote. People have to be careful to let a person like me who speak for the truth to write the prefaces, as I will cheer for the good books and trash the bad ones definitely. This one has no difference, I won't change my views for design works. Don't blame me if it's not sold well!

I take sometime travelling abroad every year, and also visit the home of many foreign friends, appreciate many awesome buildings, see a lot of great interior design. What impressed me the most is a mansion of a U.S. friend, which is more than 1,000 square meters with dense forest surrounded and refreshing indoor decorations - totally natural, and it's one of the best I've ever seen for years.

I also find some natural harmony design in this book, which is the highlight and the reason why I mark 70 points after I read through. Don't blame me, actually not many domestic books worth 60 points in my opinion.

The content of this book is very rich, but not all is fabulous. An interior designer is to make a concrete building into an exotic life works of art, and an interior design photographer is to express the interior moving lines, and the art editor is to focus on design and foil with composition of expression, only the combination of these several aspects can provide readers real enjoyment.

Serious attitude is the key to our design! Based on it and maintaining a sincere heart can ensure that walking at night, you won't be worried about encountering a ghost!

Xiao Aihua
2012.06

前言 I
不得不说

现在我每个月都要为一些设计类书籍或杂志写序或前言,写多了就有点害怕,因为序和前言是对一本书的整体概括,写得不对的时候很怕误导读者,我也知道自己很情绪化,所以我写杂文都来自内心真实的情感。自从写杂文以来,从来没违背过自己的良知和社会道义,也从来没有因为金钱而写过文章。

从七年前开始码字,到现在,我也不记得自己写过多少杂文了。现在的编辑水平越来越高,读者的眼睛也越来越亮,不写来自内心的思想是很难过关的。而现在,设计类的出版物又层出不穷,水平参差不齐,要想找到一本真正的好书也越来越难,自然我在应一些编辑邀请写序的时候就没有太多好话可说了。记得今年我就写过一篇对书评价不高的序,对图书的排版质量、设计水平说三道四,但有这么鲁莽的写序人也就有勇敢的编辑,竟然把我写的序一字不改刊登出来,这份勇气很值得我佩服,看样子我的确小看了现在的编辑。

写序和写前言的人都喜欢说好话,翻看许多的设计图书,前言和序基本千篇一律,所以我们这些做设计的看设计类书籍一般不看序和前言,直奔主题看作品,以自己的评判来判断作品的好坏,再决定是否购买。从五年前应邀开始给一些书写前言以来,有许多出版社的朋友说我写的杂文已经拥有了一批设计界的粉丝,也告知我有许多设计师是先看我写的杂文再决定是否购买此书。阿弥陀佛,在这里说一句,因为我写序或前言而使这本书不好卖的可千万别怪我,我这个家伙一直是狗嘴里吐不出象牙,要我写前言一定要悠着点,看到好的,我会鼓掌叫好,看到不好的也别怪我说话直接,今天要写这个前言同样不会改变自己对设计作品的观点,如果此书销售不好可别怪我!

每年我都会抽时间在国外游走,也去过许多国外朋友的家,看过许多很牛的建筑,也见过许多很棒的室内设计。记忆中比较深刻的是美国一个朋友的豪宅,有1000多平方米,周边被茂密的树林围绕,室内清清雅雅,与环境浑然天成,是我多年游走国外看过的最棒的住宅之一。

本书中我发现也有这种与自然环境和谐共处的设计,这也是我对这本书眼睛一亮的原因。当翻阅到后面的时候,我给这本设计类图书打上70分!不好意思,别见怪,可以入我眼的设计类图书,国内的还没有几个超过60分的!

本书作品很丰富,但并不全是精彩之作,一个室内设计师做设计的时候是把一个毛坯做成具有生活情调的艺术品,而一个拍摄室内设计的摄影师则是要把室内设计的动线感觉表达出来,一本书的美编要把设计的重点及陪衬用构图表达出来,有这几个方面相结合,才可以给读者真正美的享受。

认真才是我们设计的重点!而在认真的基础上保持一颗真诚的心,才可以晚上走夜路也不用担心遇到鬼!

萧爱华

2012.06

Preface II
Three Sentences to Digest

First of all, thanks publishers for providing us a platform to show our own projects and share with each other.

We know how important it is to have a developmental perspective. Our company started focusing on real estate projects from 2008, fortunately we become the real estate business partners in recent years gradually. Interior design is fundamentally belongs to service industry, our client is the real estate business which is the most important component of urban development of China's urbanization process, and also the most powerful push of the change.

Communication is a very important part in the process of completing the design task, and it needs a good platform to illustrate our point of view. Such as the project location, sales groups, geographical environment - all require a thorough understanding and clarity to make judgments, so that we may know the project better and try to avoid an imbalance, and then develop a complete design services program.

Each program has its own features, but I hope to achieve the design goal more from the practical use for residential projects, to focus on "people-oriented" design concept and daily life, to reflect the characteristics of the project through some of the colors, materials, etc. by finding an accurate element and a symbol to combine a kind of culture.

An excellent designer, of course, wants to make good work, but before that, I think he/she has to understand three sentences: 1. "In a party of three there must be one whom I can learn from ", which tells us that the experience of peers is valuable asset and a source of inspiration; 2. "Know yourself and your enemy well , for our own use" is to learn to grasp the advantages of other related industries, good at borrowing and making good use of reasonable sources, because the help from those industries should not be underestimated; 3. "Many hands make light work. ", no matter how good a designer could be, he can reach his limit, and a good team will always be your ladder for making progress and success.

These seem just three simple sentences, while if they can be really digested, then fame won't be far away from you.

Wang Xiaofeng
2012.06

前言 II
走在设计的路上

首先很感谢出版社给我们一个交流的平台,让我们展示各自的项目与大家分享。

我们知道,发展的眼光是很重要的。我们公司在2008年开始专注于房地产项目,很庆幸在这几年的发展中逐步成为各地产商的合作伙伴。室内设计从根本上讲是服务性行业,我们的服务对象就是房地产商。房地产是当前中国城市化进程中最重要的组成部分,是城市发展中最有力的推动者。

在完成设计任务的过程中,沟通是一个很重要的环节,需要一个良好的平台来阐明大家的观点。比如在项目定位、销售群体、地理环境等方面都需要做出透彻的理解和清晰的判断,这样对项目的认知和把握程度就会更深入,尽量避免不协调的现象,进而制定出一份完整的设计服务方案。

每个项目都有它既定的特点,但在住宅项目这方面,我希望多从实际需求上实现设计目的,重点突出"以人为本"的设计理念,把设计的重点放在日常生活当中,找准一种元素,用对一种符号,组合出一种文化,再通过一些色调、材料等来体现项目的特点。

一名出色的设计师,固然想做出优秀的作品,但是在实现之前,我认为要读懂如下三句话。1."三人行,必有我师",它告诉我们同行的经验是我们宝贵的财富,也会成为你灵感的源泉;2."知己知彼,为我所用",就是要学会抓住其他相关行业的优势,善于借用,巧妙使用,合理利用,因为这些行业对你的辅佐能力不容小觑;3."众人拾柴火焰高",再怎么优秀的设计师,他所发挥出来的光和热都是有限的,一个良好的团队永远是你进步与成功的阶梯。

这貌似简单的三句话,如果真的能把它们看懂、弄透,那么优秀的名号也就离你不远了。

王小锋
2012.06

Preface III
On the Way of Design

With this rich and information convenient time comes, the interior design industry has become more commercial and diversified with its rapid development. Open the book " Neo-classical Villas" edited by Chinese Culture, I'm impressed by the feelings brought by design - emotional, retro, fashion, and mix&match which lead us into the era of a new concept.

After a decade of dedication to my own design work, I almost have all of the emotions revolving around it - struggling, rebirth and joy. I hope my work can bring people a way of life with colorful and emotional life philosophy. What goes around comes back around. Always put yourself in others' shoes. If you feel that it hurts you, it probably hurts the others too. Leave yourself to the design and from the design, you can find a way to express yourself.

Looking back to breakdown all my works, the achievement and regret are no longer important, but what matters the most is my perseverance on the way of design, on which, I am pleased to see myself more and more sensitive to my own design with more power, perception, and still full of enthusiasm and passion for creation. In the process, I can enjoy the initial joy and subtle emotional feedbacks to continuously enrich the heart.

Chinese Culture provides a good platform, through which we might open the eyes with an open mind to learn about cases of excellent design innovation, focus on the essence of traditional and classic design sparkling for thousands of years to nourish the body and mind. Meanwhile, in the poem of "The chaotic flower gradually enchanted the eyes", it's necessary and meaningful to discard those styles by piling up blindly as to get the extreme luxury and Western style, but change to the pursuit of grandiose beauty of the luxurious nature.

Last but not least, I hope to see the collision and sparkles through the designer's perspective and communication from the limited book pages. The designers can hold the view of the whole world and pay more attention to the local traditional architecture and culture, pass on to next generation, and highly value the significance of "looking back", respectfully absorb and master the hidden traditional design philosophy behind the design work and culture, and re-examine ourselves in the culture full of vitality and endurance in the design of the buildings.

Shi Linyan
2012.06

前言 Ⅲ
走在设计的路上

在物质丰富、信息便捷的时代，室内设计行业伴随着快速的商业发展而越来越多元化，打开中华文化编辑的《以奢为美——新古典豪宅设计》一书，不禁感叹设计带给我们的强烈感触，感性的、复古的、时尚的、混搭的，百花齐放般带领着我们走进一个全新的概念时代。

经历了十年设计磨炼的我，几乎所有的喜怒哀乐都围绕着自己的设计作品展开，在里面经受挣扎、重生与欣喜。面对自己的作品，我希望传递给人们一种带有情感色彩的生活方式与生活理念。作用于人亦作用于己，己所不欲勿施于人，负责任地把自己交给设计，同时也从设计中寻找表达自己的方式。

回过头来，细数自己的作品，曾经的收获和遗憾，变得都不再重要，重要的是十年后我依然坚信着我的坚持，走在属于自己的设计道路上。在反复的摸索中，我很高兴地看到自己对设计能够越来越敏感，越来越充满发现力、感知力，依然富有不断创造的热忱与激情。在这一过程中，欣喜如最初，感受一种带有回馈意味的"情感"沙粒，从而不断地充实内心。

中华文化提供了一个很好的交流平台，通过这个平台，我们不妨以开放的心态打开双眼，了解并借鉴优秀设计案例的创新，着力于采撷那些跨越了千年仍熠熠生辉的传统与经典的设计本源，以此滋养身心。同时，在"乱花渐欲迷人眼"的风格流派中，摒弃那种将盲目堆砌作为奢华与时尚的极端走向，努力追求从浮夸的奢华向美的奢华的本质转变，这是必要而有意义的。

透过本书有限的篇幅，希望设计师的视角能有所碰撞和交流，在放眼世界的同时，更加重视本土传统建筑文化的精髓，血脉传承；更加重视"向后看"的意义所在，谦恭地吸纳和掌握隐藏在传统设计作品背后的设计思想与文化，在充满生命力与持久力的设计文化构建中重新审视自我。

史林艳
2012.06

NEO-CLASSICAL VILLAS

10	英伦风格的完美诠释 PERFECT INTERPRETATION OF BRITISH-STYLE 苏州九龙仓国宾一号别墅样板房室内设计		110	南美的乡村生活 BUCOLIC LIFE IN SOUTH AMERICA 保利阆峰云墅L18示范单位
20	美式庄园生活的返璞归真 AMERICAN ESTATE BACK TO NATURE 苏州九龙仓国宾一号别墅样板房室内设计		116	一湖碧水 一宅天下 WORLD OF CLEAR WATER AND LAKE HOUSE 维多利亚港湾别墅样板房
28	隐"墅"无形 HIDDEN "VILLA" INVISIBLE 江南一品H12		124	法式宫廷艺术的现代演绎 FRENCH ART OF THE MODERN INTERPRETATION OF THE COURT 上海奉贤法式别墅
38	人间天堂里的禅意时空 BUDDISM IN THE HEAVEN ON EARTH 中海胥江府		136	美式新古典演绎精彩人生 AMERICAN NEO-CLASSICISM MAKING GREAT LIFE 深圳中信红树湾住宅
46	中西文化的浪漫交融 THE BLEND OF CHINESE AND WESTERN CULTURE 杭州某平层公寓		144	古朴拙雅之美式风 SIMPLE AND ELEGANT AMERICAN STYLE 云宙天朗高尔夫别墅
54	中国古典文化中的浪漫主义 ROMANTICISM IN CLASSICAL CHINESE CULTURE 佘山高尔夫		152	跨越传统与时尚的现代新古典 ACROSS THE TRADITIONAL AND THE MODERN NEO-CLASSICAL STYLE 福州（融侨城）
62	佘山的梦想 SHESHAN DREAM 上海佘山三号		162	岁月如歌 TIME FLIES 福州某别墅
70	东方新贵POLY CRYSTAL HILL ORIENTAL UPSTART POLY CRYSTAL HILL 南京紫晶山A2 户型		170	庄园生活之魅 THE GLAMOUR OF LIVING IN MANOR 比华利别墅
80	艳阳高照的纯美自然 PURE NATURE UNDER THE SHINING SUN 保利阆峰云墅C1N-5示范单位		178	新欧式风格来袭 THE NEW EUROPEAN-STYLE STRIKES 长乐世纪公馆某住宅
90	后现代古典主义的装饰情怀 POST-MODERN CLASSICISM DECORATION 江门上城铂雍汇别墅C1户型		188	现代府邸里的西洋调 WEST MELODY IN MODERN HOME 玛斯兰德郭先生雅宅
100	法兰西贵族生活艺术 FRENCH ARISTOCRATIC ART OF LIVING 杭州大华·西野风韵			

Contents 前 言

194	中世纪古堡的现代演绎 MODERN INTERPRETATION OF MEDIEVAL CASTLE 皇冠钻石双层建元地产	270	新雅皮士POLY CRYSTAL HILL NEW YUPPIES POLY CRYSTAL HILL 南京紫晶山3B 户型
202	几何光影 GEOMETRIC SHADOW 温州某住宅	276	波尔多左岸的红酒 BORDEAUX WINES 波尔多左岸
208	寻访生命里最美的时光 LOOKING FOR THE BEST TIME IN LIFE 波托菲诺复式	282	古欧洲的典雅风范 THE ELEGANT ANCIENT EUROPEAN STYLE 常熟虞景山庄别墅
214	畅享优雅的钢琴曲 ENJOYING ELEGANT PIANO MUSIC 福州香榭丽舍	292	低调的新古典主义 UNDERSTATED NEO-CLASSICISM 波托菲诺复式住宅
220	豪宅新主张 MODERN MINDS ON VILLA 翠湖绿洲花园示范单位	298	蜘蛛侠与爵士乐 SPIDER MAN AND JAZZ 嘉宏湾花园样板房
228	热情而斑斓的泰式风情 ZEALOUS AND GORGEOUS THAI STYLE 中颐永和（A401单元联排别墅）	304	奏响浪漫的华美乐章 PLAY THE BEAUTIFUL ROMANTIC MUSIC 深圳大东城·嘉宏湾花园
236	月亮之于夜空的空间精神 THE SPACIAL SPIRIT WHEN THE MOON IN THE SKY 珠海中信红树湾28-9别墅示范单位	312	奢华与浪漫的新符号 NEW DEFINITION OF LUXURY AND ROMANCE 安徽华地公馆B1-1样板房
244	禅韵·江南风采 ZEN RHYME. JIANGNAN STYLE 福州某单层住宅	318	关于空间 关于美 WITH REGARD TO SPACE AND BEAUTY 简欧风格·珀丽湾别墅
250	梦回明清 DREAMS OF MING AND QING DYNASTIES 南山清水溪别墅	326	摩登时代 MODERN TIMES 福州某摩登住宅
258	中西精粹于一墅 THE VILLA COMBINED THE EAST AND WEST TOGETHER 古北佘山国际别墅	330	都市新贵 URBAN UPSTART 某私人住宅
264	墨情空间 INK PAINTING SPACE 福州某住宅		

PERFECT INTERPRETATION OF BRITISH-STYLE

英伦风格的完美诠释

项目名称：苏州九龙仓国宾一号别墅样板房室内设计
设计公司：上海大研室内设计工程有限公司
设 计 师：欧阳辉
项目地点：苏州
项目面积：601 m²
主要材料：黑金花石材、摩卡绿、波曼米黄、石材拼花、马赛克等

Project Name: Suzhou Kowloon Warehouse Ambassador Number One Villa Model Room Interior Design
Design Company: Shanghai Dayan Interior Design Institute and Engineering Co., Ltd.
Designer: Ouyang Hui
Project Location: Suzhou
Project Area: 601 m²
Main Materials: black golden flower stone, mocha green, Portman beige, stone mosaic, mosaic, etc.

以奢为美 | 新古典豪宅设计

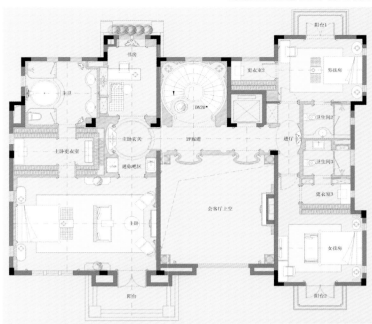

英伦风格体现出的是一种典雅庄重，兼具绅士风度的住宅空间，业主的传统素养、严谨的做派在此套住宅中得到了完美的诠释。
步入室内，一条主线贯穿玄关、厨房、餐厅、客厅。深色木墙与米黄色皮革软包相互搭配，咖啡色系的花纹布艺、沙发、地毯丰富了室内的生活气息，沉稳而不古板。

British style reflects a kind of elegant and dignified residential space as a gentleman. The house is a perfect interpretation of the owners' traditional literacy, rigorous character. Coming into the indoors, people can find a main theme from the entrance, kitchen, dining room, to the living room. The dark wooden walls and beige soft leather bag matches with each other very well. The brown pattern fabric, sofa, carpet enriches the living atmosphere, which is calm and not stuffy.

二层是私密休息区，宽敞的卧室增强了起居室的功能，男主人一早会坐在这里看报纸，而女主人则会在更衣室精心挑选她的服装。双台盆可供两人同时洗漱，此层还设有主卧和书房、儿童房。双独立套间的配备，每个区域都有着各自的使用空间，使生活更便利、更舒心。

地下室中有工作室以及保姆房、娱乐室。娱乐室的正中央摆放着一张台球桌，这项高雅休闲的运动深受绅士们的喜爱，可邀好友来此切磋球技，同时也聊些打球之外的话题。独立收藏室和吧台是男主人的最爱，男主人可以邀请关系亲密的三五好友在此分享自己珍贵的藏品，一起探讨收藏的乐趣，同时品味自己收藏多年的美酒，更展现出男主人的个人魅力和生活品质。

The second floor is a private lounge area, and the spacious living room enhances, functions of the bedroom. Master can be sitting here reading the newspaper early in the morning, and the hostess can be carefully selected her clothing from the dressing room. Double wash basin can be used for two at the same time. A master bedroom and study, children's room are also set on this floor. Equipped with dual independent suites, each region has its own use of space, so that people can live more convenient and comfortable in their daily life.

A workshop, a nanny room, and a recreation room are in the basement. A pool table is placed in the center of the entertainment room, as this activity is loved by the elegant gentlemen, and they will come to demonstrate their skills with friends, while playing they also chat about the topics beyond pool. Separate collection room and the bar are the master' favorite, and he can invite a few close friends here to share his precious collections and explore fun of collection, and taste his wine collection from many years ago - all of which can show the owner's personal charm and quality of life.

AMERICAN ESTATE BACK TO NATURE

美式庄园生活的返璞归真

项目名称：苏州九龙仓国宾一号别墅样板房室内设计
设计公司：上海大研室内设计工程有限公司
设 计 师：欧阳辉
项目地点：苏州
建筑面积：375 m²
主要材料：天然石材、仿古砖

Project Name: Suzhou kowloon Warehouse Number One Villa Ambassador Model Room Interior Design
Design Company: Shanghai Dayan Interior Design Institute and Engineering Co., Ltd.
Designer: Ouyang Hui
Project Location: Suzhou
Project Area: 375 m²
Main Materials: natural stone, antique brick

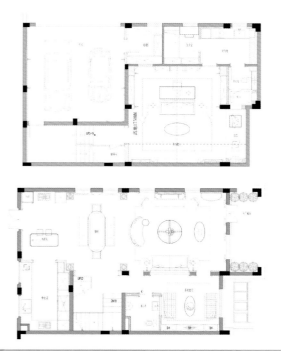

该案位于苏州工业园区金鸡湖大道旁，位置非常理想。此套样板房定位于美式古典风格。在室内环境中力求表现悠闲、舒畅的生活状态。运用了天然石材、仿古砖等石材来体现业主优雅的生活追求。

一层设有客厅、餐厅和厨房。中、西式厨房的巧妙结合，满足了主人对美食的热爱以及对生活品质的追求。客厅与餐厅中，纯色铁艺的材质体现出美式风格的粗犷与自然感。

二层是主人的私密空间。在美国人的价值观念中，卧室应该是最豪华的地方。本案中，进门是独立的书房，西边是主人的卧室、更衣间和卫生间。整个空间的线条流畅且功能统一。

三层是儿童房。男女小朋友都有自己的主题生活空间和独立的活动区域。

地下室部分主要设置了视听室、书房等多功能空间，体现出主人对生活品质的追求。

该案给人以原始而简洁、粗犷而随性的感觉，满足了人们返璞归真的心理需求。

This apartment is located besides Jinji Lake of Suzhou Industrial Park Avenue, which is quite ideal. This set of model room is designed in American classical style. The indoor environment is designed to be with relaxing and comfortable living conditions. The use of natural stone, antique tiles and other stones reflects the owners' pursuit of the elegant life.

There are living room, dining room and kitchen on the first floor of the villa. The unique combination of Western-style and Chinese-style kitchen meets the owner's love for food and the pursuit of quality of life. The solid iron material in the living room and dining room reflects the rough and natural quality of American style.

Second floor is the master's private space. In American values, the bedroom should be the most luxurious place. In this apartment, an independent study is designed right facing the door, and the master bedroom, dressing room and bathroom are to the west. The entire apartment is with smooth lines and unified features.

The kid's room is on the third floor. The boy and girl have their own separate theme living space and activity areas.

The auditorium, study and other multi-functional space are set in the basement, which reflects the owner's pursuit of quality life.

The design presents a primitive and simple, rugged and casual style, which meets people's psychological needs for coming back to nature.

HIDDEN "VILLA" INVISIBLE

隐"墅"无形

项目名称：江南一品H12
设计单位：宁波神采装饰设计工程有限公司
设 计 师：史林艳
项目面积：400 m²
主要材料：樱桃木饰面、西班牙米黄石材、帝王金石材、黑金花石材

Project Name: Jiangnan Yipin H12
Design Company: Ningbo Shencai Decoration Design Engineering Co. Ltd.
Designer: Shi Linyan
Project Area: 400 m²
Main Materials: cherry veneer, Spain beige stone, imperial gold stone, black gold flower stone

■ 以奢为美 ▎新古典豪宅设计

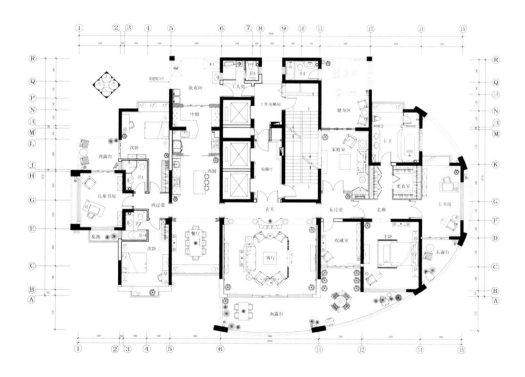

江南一品H12户型，按每层约400平方米进行单层独立式设计，在室内布局上秉承一般别墅都具有的功能规划。整体外观似乎与一般性高层公寓住宅无异，但一户一层、各自独立的设计构想恰如其分地满足了业主奢华而不夸耀、隐逸而不张扬的心理追求。在室内设计上，将六分贵气、三分静气、一分清气融为一体，体现了美式风格低调奢华的理念。

大量米黄色大理石和金黄色石材的运用，奠定了整部作品富丽堂皇的基调。玄关的凹形石材立面造型、大理石地面拼花，走廊的四根罗马式柱、背景立面的石材边框等无不在步入客厅的第一时间让你感知一种庄严大气。沙发、茶几、衣柜、床、书柜、书桌等多采用美式风格，用料考究，注重细节。局部以花纹雕刻点缀，以褐色、红橡木原色为基本色。各类水晶吊灯的运用，与整体风格浑然一体，更显璀璨炫目。而独居客厅一隅的老式留声机，默默无声地将一种遥远的品位感、历史积淀感缓缓吐露，更是一种身份地位的象征。

有才而性缓定属大才，有智而气和斯为大智。做奢华易，做奢华而不张扬难。如何将奢华做到低调而沉稳，就需要在"性缓"和"气和"上下功夫。整部作品强调色彩的和谐与呼应，如客厅镶嵌金边的天花与地板石材色的呼应，壁炉上方香槟金色的圆镜与家具色彩的呼应，均从细节处实现了一种协调统一。在各种家具和装饰的摆设上，注重从不同角度和侧面去诠释主题，避免堆砌和冲突，在和谐的氛围中将三分静气展现无遗。

为实现沉稳而不沉闷，就需要有一分清丽之气来衬托。从设计上来看，设计师采用少量绿色大叶植物装点各个空间，给室内增添几分自然和活力；采用局部的跳色，使色彩不过于单调，如客厅壁炉使用白色大理石和黑白花纹拼花，旁边的两把椅子则使用黑白条纹坐垫与之呼应，使人眼前一亮而不觉突兀；男女儿童房则依据需要使用"性格色彩"装饰，使整体风格不失自由、各取所需；书房内陈设简易高尔夫场地，使学习与运动两不误；观景露台能充分利用空间，使用大量绿色植物盆景，摆设咖啡桌椅，实为亲近自然、午后小憩的人工氧吧。

大音稀声，大隐于市，隐"墅"无形，江南一品H12单梯单户平层户型设计，从一定程度上将出世与入世的哲学思想做到了有机统一。

H12 unit of Jiangnan Yipin is designed as about 400 square meters for each single floor, the interior layout of the villas has a general function of planning. The overall appearance is almost the same as general apartments, but every apartment is of independent design, which appropriately meets the owner's pursuit of low key luxury life style. The interior design is combined of six richness, three quietness, and one freshness, all of which reflect the concept of American-style low-key luxury.

A lot of beige marble and gold stone make the whole design a magnificent style. The concave shape of the stone facade at the entrance, the marble parquet, the four Roman pillars in the corridor, and stone frame of the background facade, etc., all of which create solemn atmosphere when people enter the living room the first time. Sofa, teapoy, wardrobe, bed, bookcase, desk, etc. are all in American-style with refined materials and detail oriented, decorated with carved patterns for some details with brown, red oak as the basic colors. The use of various types of crystal chandeliers matches with the overall style perfectly, very gorgeous. The old-fashioned phonograph in the corner of living room reveals the history of accumulation in a silent way, displays a distant sense of taste, sense of slowness, but also a symbol of social status.

Gifted with great temper is talent, wisdom with mild temper is wise. It's easy to be luxurious, but it's hard to be so without publicity. To achieve low-key luxury you need to work on "great temper" and "mild temper". The whole works emphasizes the harmony and match of colors, such as the living room ceiling and the floor mosaic stone color echoes, champagne golden round mirror above the fireplace and furniture color echoes, which reveals the consistence of the details. The layout of different furniture and decorative furnishings focuses on the interpretation of the theme in different ways, avoids piling up and conflicts, but displays the quietness in a harmonious atmosphere.

To achieve calm but not boring effect, you need to have some elegant flavor embedded. From the design point of view, the designer uses a small amount of green leaf plants to decorate each room, thus adding a bit of nature and vitality; The use of partial jumpy color makes the overall tone not monotonous, such as in the living room fireplace with white marble and black and white pattern parquet echoes to the two chairs with black and white striped cushion, which is definitely a highlight but not too loud. The baby boy's and girl's rooms are designed based on "personality color" decoration, which makes the overall style free and practical; A simple golf course is set in study, so it makes learning and fun together; viewing terrace takes full advantage of space - uses a lot of potted green plants, decorative coffee tables and chairs, in fact all of them make people close to nature and people can enjoy the afternoon nap in this artificial oxygen bar.

Great sound is soundless; great is hidden in the city. Hidden "villa" in a city is just as Jiangnan Yipin H12 single drawer units, which unifies philosophy and living style perfectly together.

■以奢为美 ▌新古典豪宅设计

37

BUDDISM IN THE HEAVEN ON EARTH

人间天堂里的禅意时空

项目名称：中海胥江府
设计公司：HSD水平线空间设计
总设计师：琚宾
设计团队：韦金晶、谭琼妹、邱建军、石燕、尹芮
项目地点：苏州
项目面积：337 m²
主要材料：伊朗玉石、艺术肌理漆、黑钢、铁刀木、西班牙米黄石
摄 影 师：孙翔宇

Project Name: Zhonghai Xujiang Mansion
Design Company: HSD Spatial Art
Designer: Jubin
Design Team: Wei Jinjing, Tan Qiongmei, Qiu Jianjun, Shiyan, Yinrui
Project Location: Suzhou
Project Area: 337 m²
Main Materials: Iran jade, art texture paint, black steel, venge, Spanish beige marble
Photographer : Sun Xiangyu

以奢为美 新古典豪宅设计

39

以奢为美 | 新古典豪宅设计

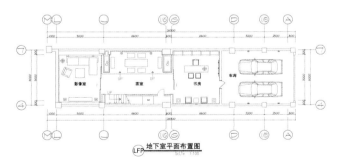
地下室平面布置图

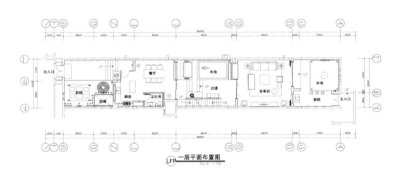
一层平面布置图

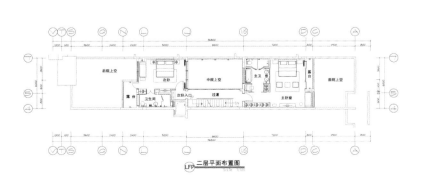
二层平面布置图

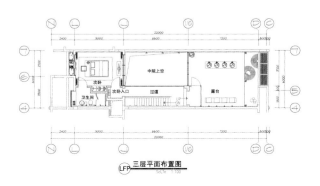
三层平面布置图

该案是设计师琚宾为中海胥江府设计的别墅，该项目坐落在素有"人间天堂"之称的苏州，在江南蒙蒙烟雨的熏染下，整个设计与悠然宁静的城市氛围完美融合。在大尺度空间中塑造出了温暖舒适的人居环境，寓自然文化于一室。设计手法简洁干净，但随处都流淌着设计者还现象于本真、道法自然的设计理念。

物质与精神

该案设计的核心是纵贯上下四层空间的室内墙面，统一的米色大理石材质，没有任何装饰的白色吊顶，剔除了多余的元素、色彩、形状和纹理，高达四层的空间里，光洁的大理石向上、向下延展开来，紧紧包裹和依附在建筑表面，创造了四白落地，如同中国画留白一样的视觉效果，从而凸显出了其他设计内容，使其成为视觉的焦点。方形黑色窗框、悠长的走廊扶手、硬朗的直线条家具，甚至矮几上的深色陶罐，都好像挥洒在白色的宣纸上，饱蘸浓墨的一横、一竖、一点，人文气息跃然纸上。

设计师通过对材料的极致运用，在物质的层面将空间结构一览无余地呈现出来，用建筑自身的力量来震撼观者的感官，将物体形态的通俗表象，凝练成为一种高度概括的抽象形式。这是一种文化提炼，是摒弃尘俗与浮华，直至本原，诚如禅悟所求，要人摆脱千般计较，在万象之中直觉体悟生命的原本面目，从而创造宁静温暖的家居氛围，使居住者达到精神的圆满与永恒，这也正是设计者所理解的，在设计中追求物质与精神的融会贯通。

共性与个性

在该案的设计中，设计者更注重空间共性的表达，从尊重建筑内在的精神入手，运用具有一致性和规律性的表达手法来处理空间的体块关系、收口关系、光影关系，利用纵向的立面设计，大块面的墙面铺装和简洁的家具陈设，挖掘建筑自身的魅力，还原居住空间——"家"的本真概念，这是对空间共性的探求。而后，设计师又根据空间自身的背景和场所感，通过对比、协调、统一等设计手法进行二次创作，通过家具风格的定位、配饰的选择和装饰材料的质感表达，利用这些显性的元素来塑造属于这一空间的独特气质。例如，餐厅墙面整块的黑色原木装饰、卧室隔断、书房中直线条网格状的粗犷书架、四边方正的书桌，这些简练硬朗的直线条家具与大体块的空间布局，在纵向和横向上相互穿插，构架出风格统一的硬质空间。而卧室内朴素、典雅的织物和木质家具渗透出的自然质朴、粗犷原始，色彩上素色调的强烈对比又迎合了空间宁静、简素的人文家居风格。这些设计元素密不可分，与空间浑然一体，使空间装饰和设计具有了唯一性。

Zhonghai Xujiang Mansion, designed by Ju Bin, is located in Suzhou, a city which is called as Heaven on Earth. In South China where rain occupies most of the days, this design presents you the tranquility within the city. In the big space, it is a pleasing place to live in with the harmony of nature and culture. The design is clear and brief, flowing an idea that is to bring all things back to nature.

Material and Spirit

This case focuses on the internal walls within the 4-storey space. Cream-colored marble material and white suspended ceiling are used, no additional elements like lines, colors, or shapes. In the whole space, the marbles spread tightly on all directions of the room, as if a traditional Chinese painting visualized in the space. This basic clean design props other designing elements like, rectangle black window frames, handrails on the long corridor, sharp lined furniture, and deep-colored pottery on the tea table. All these make a cultural environment for the space.

The designer optimizes all the materials to demonstrate the special structure. The visitors would be affected by the building itself, more importantly, this case is highly abstracted from all the things one can touch. It shows essence of culture, rejecting the vanity to restore things back to what they were. As advocated by Buddhism, one should be get rid of all the conflicts and desires, only in this way will one be able to experience the real nature of things concealed deeply behind. Tranquility and warmth of the family fill the house, and they fulfill family members with completeness of life for good, which is fully understood by the designer who tries to balance material and spirit under the same roof.

General Character and Special Character

This design expresses general character of space. The balance between the light and shadow, displays and visualization of the space are properly dealt with accordance and order, guided by the spirit of the original building. Simple-lined furniture and the marble walls show the charm of the building and the sense of home. The designer makes the second round of designing based on the previous space, by contrasting, according, and balancing the elements in the space. Decorations and other issues are used to reinforce the special character of the space like black wood on the wall of the dining room, separation of the bedroom and latticed book shelf, sharp cornered desk. Those lines work well with all the space, giving an expression of unity of the whole house. The plain and elegant decorations like fabric and wooden furniture, they exude a sense of nature, easiness and original spirit. The strong contrast between colors makes brief and cultural environment. All the elements are as one, and the whole decorations and design are unique.

一个好的设计，是物质与精神的融合，是共性与个性的共存。在千年姑苏的小桥流水中，这个设计用简洁、有序的外显特征塑造了宁静致远的空间灵魂，回应了现代生活的功能需要，丰富、深邃的内涵感悟满足了现代人的精神需求。正如墨西哥设计师路易斯·巴拉干所说："没有实现宁静的建筑师，在他精神层次的创造中是失败的。现在的建筑物不仅缺乏静谧、静默、亲切和惊奇这类概念，连美丽、灵感、魔力、魅力、神奇这类词汇也消失了，而所有这些才是我心灵的渴求。"

A good design balances material and spirit, general character and special character. With poetic image of the South China flowing thousands of years, this case provides you with what you are seeking for tranquility in eternity. It is conciseness that never fails to meet the need of modern life, and the deep meaning in the design fulfills the need of people spiritually. As the famous Mexican architect Luis Barrangan put it, "an architect who fails to translate sense of tranquility fails in design spiritually. Modern architecture are lack of tranquility, silence, affability and surprise, let alone the beauty, inspiration, magic and charm. What they lack are what I am seeking for".

THE BLEND OF CHINESE AND WESTERN CULTURE

中西文化的浪漫交融

项目名称：杭州某平层公寓
设计单位：杭州博洛尼装饰工程有限公司
设 计 师：梁苏杭
项目地点：杭州
项目面积：320 m²

Project Name: An Apartment in Hangzhou
Design Company: Hangzhou Bo Loni Decoration Engineering Co., Ltd
Designer: Liang Suhang
Project Location: Hangzhou
Project Area: 320 m²

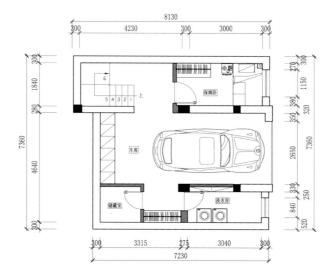
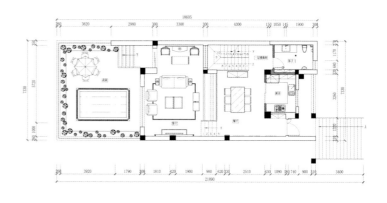

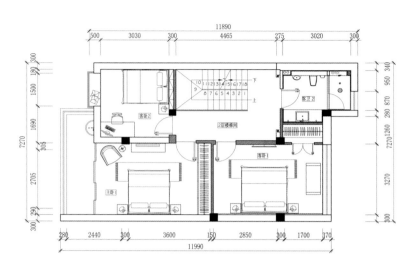
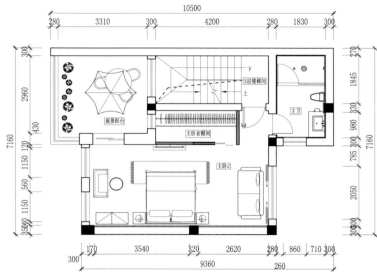

50

该案的定位为新古典——港派简约风格，这种风格吸纳了中式传统居住空间的居中、对称、均衡的格局特点，又有西方文化的唯美意境。古典饰品优雅地呈现在空间各处，设计师在细节上的点缀，体现了对以人为本理念的尊崇。非常符合现代人们的人居理念。

新古典主义兴起于18世纪的罗马，其宗旨是对巴洛克和洛可可的反对，其次以重振古希腊、古罗马的艺术为信念……它排斥对以往文化无限制的复制和无条件的接纳，提倡采用新的工艺、新的材料、新的设计理念去表现内心对美的感悟，在注重装饰效果的同时，用现代的手法和工艺还原古典气质。新古典主义具备了古典与现代的双重审美效果，其完美的结合也让人们在享受物质文明的同时也得到了精神上的慰藉。

This case is designed with a neoclassical style - a construction style with HongKong features, simple but comfortable. This kind of style assimilates the good construction features of traditional Chinese buildings-centered, symmetrical and balanced accompanying with western characteristics. The designer properly decorates the classical ornaments in this space, which reflects his person-centered design concept. Nowadays, this kind of design style is in line with people' living concept.

Neoclassicism sprang up in Rome in 18th and aimed at getting rid of Baroque and Rococo, then to revive the ancient Greek and ancient Rome style...Neoclassicism advocates the application of new process, new material and new ideas on design to show the understanding to beauty while being against the limitless copy of culture and unconditional acceptance of culture. Designer reproduces the classic beauty with modern methods and neoclassicism contains the modernism and classicism at the same time, which serves scrumptious nourishment for mind.

ROMANTICISM IN CLASSICAL CHINESE CULTURE

中国古典文化中的浪漫主义

项目名称：佘山高尔夫
设计单位：萧氏设计
主创设计：萧爱华
参与设计：黄鑫
软装设计：郭丽丽
项目面积：480 m²
摄 影 师：萧爱华

Project Name: Sheshan Golf
Design Company: Xiao's design
Head Designer: Xiao Aihua
Co-designer: Huang Xin
Soft Decoration Design: Guo Lili
Project Area: 480 m²
Photographer: Xiao Aihua

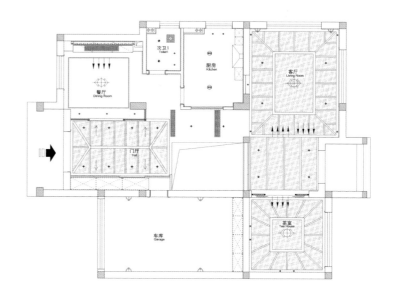

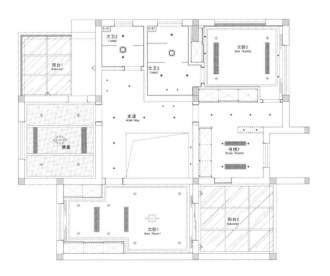

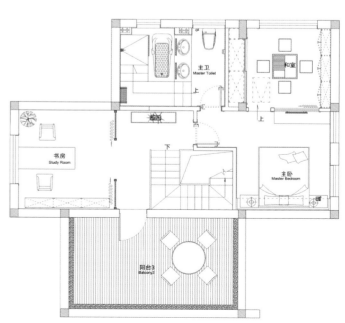

该套方案位于上海市的高端社区——佘山，业主是一对具有中国情结的美籍华人。在该案的设计中，设计师用现代时尚的方式来诠释整个空间。主体的装饰元素为树的形态，将树的形象用抽象的方式运用在一楼的各处以及卧室空间中。天棚上及墙面的藤编工艺，使空间中传达出一种自然的属性。

在软装配饰上，设计师追求独一无二的特性，其中有许多名家的作品参与其中，墙面上的挂画基本都是原创作品。在楼梯的空白处，设计师特意从古玩市场上寻找有历史痕迹的家具，所以这些都可以把作品的自然回归诠释得更加完美。

该案的空间要用360度的视角来进行全方位的审视，才能探索其中的艺术性，有时候会发现摄影机的镜头只能表现一个局部，设计师从女主人喜欢中国古典文化的浪漫主义来切入，在色彩设计上用了很多黑、白对比，同时加以色彩的点缀，从而提供了一个完美的视觉享受。

The project is located in Shanghai's elite community - Sheshan, the owners are Chinese-Americans with passion for Chinese culture. In this design, the designers use modern and stylish way to interpret the entire apartment. The main decorative elements are in tree form with the image of the tree used in the abstract way throughout the first floor and bedroom. The decorative canes on ceiling and wall are like blessing from nature.

For the soft decorative accessories, the designers pursuit the unique features - involved many famous works, almost all paintings on the wall are original works. They specifically found furniture with traces of the history from the antique market and put them on the stairs, which make perfect interpretation of the natural return.

The artistic beauty can be found from a 360-degree perspective, sometimes the camera lens can only capture the part of the design effect. The designer took Chinese classical culture from Romanticism the hostess preferred to cut in - some contrast is used in the color design with a lot of black and white to show a perfect visual effect.

以奢为美 新古典豪宅设计

61

SHESHAN DREAM

佘山的梦想

项目名称：上海佘山三号
设计公司：萧氏设计
设 计 师：萧爱彬
配合设计：屠江江
陈设设计：郭丽丽
项目地点：上海
项目面积：380 m²
主要材料：原木地板、香草米黄大理石
家具饰品：递展家居（意大利）、法视界家具、当代艺术家作品
摄 影 师：萧爱华

Project Name: Shanghai Sheshan No.3
Design Company: Xiao's Design
Designer: Xiao Aibin
Co-designer: Tu Jiangjiang
Furnishings Design: Guo Lili
Project Location: Shanghai
Project Area: 380 m²
Main Materials: wood floors, vanilla beige marble
Furniture Accessories: home delivery exhibition (Italy), France horizon furniture, works of contemporary artists
Photographer: Xiao Aihua

佘山三号是上海繁华都市的一块净土,浓密的绿化掩盖着栋栋别墅。略显禅意的空间能令每个观者马上安静下来,门厅中薛松的画作《毛泽东》顿使人们感受到空间的不俗气息。门及门框高挑、挺拔,客厅的高度,餐厅的宽度,给人一种犹如步入园林的感觉。空间随着你的移动,有节奏地变换着。当你静下心来细细品味的时候,你会发现设计师在做设计的时候是经过精心安排的,在空间的转换上费尽心思。萧氏设计一直以来很重视空间心理学的研究,希望能够设计出最让人舒服和愉悦的空间,视觉的享受和功能的完善是最重要的部分,然后是材质和工艺的选择。该案就把这四个要点完美地呈现出来。

此外,灯具是设计师为此空间量身定制的,画作和雕塑都是业主的收藏,前卫艺术家陈志光的"蚂蚁"令空间气度不凡,楼道照片的主题是建筑大师密斯·范德罗设计的巴塞罗那馆,从中可以看出业主的喜好和对艺术的追求。

Sheshan NO.3 was built on a pure land in the bustling city of Shanghai - villas were set in green nature one after another. The design with some Zen elements makes every visitor calm down immediately, and the elegance can be found right through Xue Song's painting "Mao" hanging in the hall. The doors and frames are particularly tall and straight, the height of living room and the width of the dining room make it feel like wondering in the garden. Space is changing as you move with the rhythm. When you slow down and feel it by heart, you will find it was delicately refined by the designer, especially spent some effort on the space conversion. Xiao's Design always attach great importance to the design of space psychology research to present the most pleasant, comfortable and pleasant design. The visual and functional improvement is the most important part, and then comes the materials and processes. This design presents all the four elements perfectly.

In addition, the lighting is tailored for this case by the designer, and paintings and sculptures are the owner's collection, the work "ants" from modern artists Chen Zhiguang makes the house very special. The corridor theme is the Barcelona Pavilion designed by architect Ludwing Mies Vander Rohe, which reveals the owner's preferences and the pursuit of art.

ORIENTAL UPSTART POLY CRYSTAL HILL

东方新贵POLY CRYSTAL HILL

项目名称：南京紫晶山A2 户型
设计单位：广州尚逸装饰设计有限公司
设 计 师：王赟、王小锋
投 资 商：保利地产
项目面积：188 m²

Project Name: Nan Jing Purple Crystal Hill A2
Design Company: Guangzhou Shangyi Decoration Design Co., Ltd.
Designer: Wang Bin, Wang Xiaofeng
Invest Company: Poly Real Estate
Project Area: 188 m²

该案是对传统中式风格的贵族化复兴，设计师把中国古代帝王将相的生活空间，用现代的处理手法打造成既实用又不失雍容典雅意味的生活空间。旨在打造中国文化精髓的同时，吸纳西方美学的精华，利用空间穿插和色调渲染缔造出具有浓郁中式情调的生活空间。

在空间布局上，设计师最大限度地提高利用率，在满足生活需要的同时，保持阳光的渗透和空气的流畅。空间塑造上体现出中式园林的"移步换景"、"框景"、"对景"法则，古朴中求新颖。

在公共空间中，地面主要采用了天然大理石，立面是麻质墙纸与实木饰面，结合横竖纹拼接等特殊处理工艺，中情西韵、亮丽华贵。在配饰风格上，高贵恒久的紫檀家具、宫灯、花鸟和景泰蓝都是中式的核心所在，在家具和摆设的搭配上，设计师巧妙地把古典和现代、深沉与轻快、古朴与跳跃结合了起来。

This project aims at renovating the traditional Chinese style to make the aristocratic feature outstanding. Designer creates an imperial life space with modern methods. When to be compatible with aesthetic characteristics both Chinese Culture and Western culture, designer creates fully-Chinese life space by inserting space and smudging color.

In terms of space arrangement, designer makes every endeavor to take full advantages of the whole space to let the sunshine come into the room and ensure the air flow under the circumstances that every requirement is satisfied. Designer applies many Chinese garden principles to shape the space. In this design, 'scenery changing in every step' 'scenery unframed' and 'opposite scenery' these Chinese principles are applied perfectly.

The main material for floor of public space is crude marble, and the material for wall is the combination of linen wall paper and wooden over coating, which shows both Chinese culture and Western culture with the special methods in dealing with the pattern design. In terms of decorations, red sandalwood furniture, painting of flowers and birds, palace lantern and cloisonné play a key role in showing the Chinese characteristics. Designer combines not only classic and modern style together, but the feeling of prudent and lively together, and permeates them into the furniture setting.

以奢为美 新古典豪宅设计

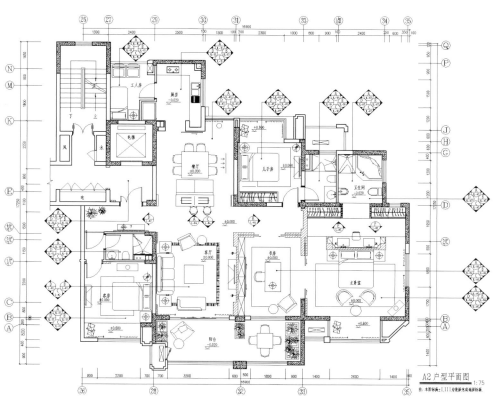

A2户型平面图

PURE NATURE UNDER THE SHINING SUN

艳阳高照的纯美自然

项目名称：保利阆峰云墅C1N-5示范单位
设计单位：广州市韦格斯杨设计有限公司
项目地点：长沙
建筑面积：472 m²
主要材料：米黄/深啡网大理石、马赛克、印花瓷砖、涂料、木饰面、木地板、铁艺、墙纸

Project Name: Poly Langyunfeng C1N-5 Showflat
Design Company: GrandGhostCanyon Designers Associates Ltd.
Project Location: Changsha
Project Area : 472 m²
Main Materials: beige/dark emperador marble, mosaic, printing ceramic tile, paint, wood finishes, wooden floor, iron, wallpaper

以奢为美 新古典豪宅设计

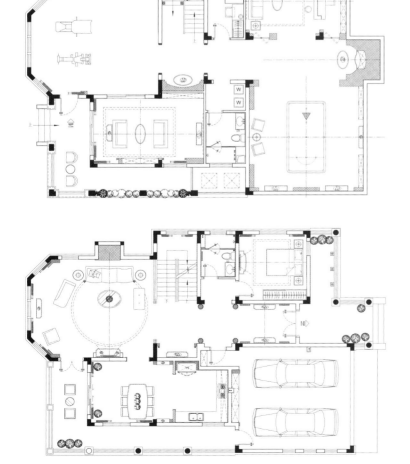

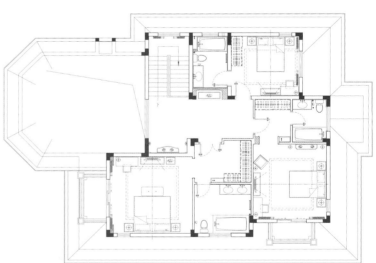

该案以蓝、白两色为空间的主色调,搭配原木色家具及柔和色调的布艺,使人感受到一种地中海蔚蓝色的浪漫情怀,在艳阳高照的纯美自然中,使人心情变得轻松而平静。

该案运用连续的拱门分隔空间,挑空大厅的墙身材料取材于大自然,选用了白橡木刷白漆,流畅的几何线形分缝,搭配明快的白色橡木,橡木所独有的纹理,极具自然的气息。大面积的铁艺装饰配合天花的大型铁艺吊灯以及墙身的铁艺壁灯,流露出古老的气息,小巧可爱的绿色盆栽及鲜花也给空间增加了不少活力。

餐厅墙身也是运用肌理漆,随意而极具感染力的纹理搭配半凿的拱形和蓝、白相间的马赛克拼花,在仿古砖及小花砖的映衬下,更加优雅脱俗。

穿过拱形的门洞,映入眼帘的是木结构的造型柱,拾级而上是实木扶手及特殊工艺打造的铁艺栏杆,在铁艺水晶灯的照耀下显得非常有趣。

主人房采用了白色橡木搭配清爽的墙纸,木饰面也是运用了流畅的竖向分缝,使其与其他空间自然地连为一体,主幅的弧形造型及布艺挂饰是整个空间的视觉重点,设计师巧妙地将原建筑中不对称的顶棚变为方正对称的造型,衬以花纹墙纸,在造型考究的铁艺吊灯柔和灯光的照射下,烘托出一种安静、浪漫的氛围。

白色的木饰面、肌理漆的感染力、石材的纹理与光泽、铁艺的独特造型、色彩鲜明的家具、充满生命力的植物,所有这一切,都使我们感受到古老而又不失现代气息的地中海风情。

White and blue constitutes the main color of the project. The wood-color furniture and soft color cloth art issues create a romantic feeling as if you were in the Mediterranean Sea. Under the shining sun, you shall enjoy the easiness and peace.

In this case, arch doors separate the space, and walls of living room with higher ceilings are modeled on natural materials. White oak paintings on the wall with smooth lines give you a sense of going back to nature. Iron-made decorations go well with the iron chandelier and wall lamps, exuding anciently. Lovely bonsai and flowers bring life to the space.

The walls in the dining hall are painted of natural texture paint. Casual grains, half-chiseled arch, blue and white mosaic make the old-style bricks and flower bricks graceful.

Through the arch doors come the wooden pillars. When you take the stairs, you can see the special iron banisters and wooden handrails, which is so interesting with the lights of the iron crystal lights.

The bedroom is decorated with white oak and refreshing wallpapers and vertical parting is applied to the wooden board. All these connect well to make the decoration as a whole. The main arch shape and cloth issues are the center of the space. The designers changed the asymmetrical ceiling a balanced one, and the light from the iron lights projects a sense of romance and tranquility. The white wooden board, the impressive natural texture painting, and the grains of the stone material, unique iron-made styles, bright-color furniture, and green plants… all these demonstrate the beauty of the ancient and modern Mediterranean style.

POST-MODERN CLASSICISM DECORATION

后现代古典主义的装饰情怀

项目名称：江门上城铂雍汇别墅C1户型
设计单位：J2-DESIGN顾问设计有限公司
主设计师：苏挚邦/设计F组
陈设设计：区婷婷/陈设材料C1组
材料设计：徐流艳/陈设材料C1组
项目地点：江门
项目面积：480 m²
主要材料：大花白、黑白根、深啡网、浅啡网、黄金海岸等石材、玫瑰图案马赛克、银箔纸等

Project Name: Jiangmen Shangcheng Boyonghui Villa C1 Type of Flat
Design Company: J2-DESIGN Consulting and Design Co., Ltd
Chief Designer: Su Zhibang in Design Group F
Display Designer: Qu Tingting in Display Material Design Group C1
Material Design: Xu Liuyan in Display Material Design Group C1
Project Location: Jiangmen
Project Area: 480 m²
Main Materials: Arabescato, Nero Margiua, Deep Coffee Network Marble, Light Coffee Network marble, Golden Bay marble and other stone, rose pattern mosaic, silver foil paper, etc.

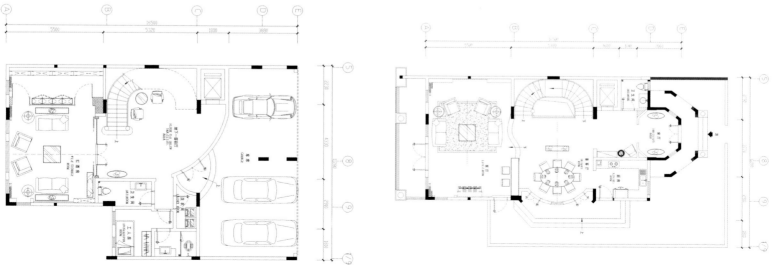

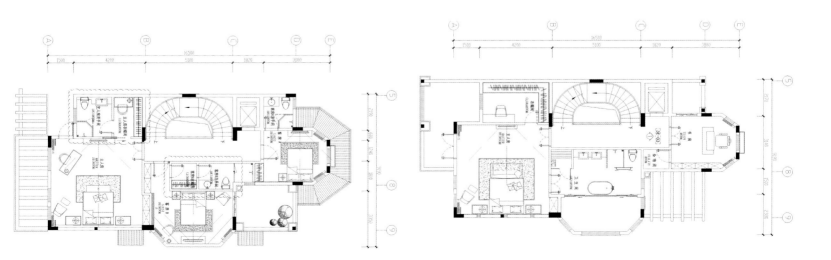

该案设计注重线条的搭配和颜色对于空间气氛的营造，在造型设计的构图理论中吸取其他艺术或自然科学的概念，把传统的构件通过重新组合寓情于景，追求品位与色彩搭配的和谐。大厅中运用大量的白色元素，现代感十足，又不失低调奢华与柔美的感觉；二层房间运用淡雅的色调，不张扬的美充满了温馨感，但从低调的氛围中却使人感受到了贵气十足的韵味。

In this case, designers attach importance to the collocation of lines and use color to create spacial atmosphere. The composition theory in design takes in concepts in art or natural science, through reorganization of traditional elements, the design pursues harmony between taste and color matching, expressing ideal in composition. White is widely used in lobby, which makes the lobby rich in modern sense but also full of low key luxury and gentle beauty; on the second floor the quietly elegant color and the low key beauty are full of warm sense, but from the low key atmosphere, people can sense rich royalty in it.

FRENCH ARISTOCRATIC ART OF LIVING

法兰西贵族生活艺术

项目名称：杭州大华·西野风韵
设计单位：深圳市世纪雅典居装饰设计工程有限公司
设 计 师：聂剑平、蒋禄川
项目地点：杭州
项目面积：840 m²
主要材料：大理石、橡木、墙布、墙纸、质感涂料、艺术玻璃

Project Name: Hangzhou Da Hua Wild West Charm
Design Company: Shenzhen Athena Century Decoration Design Engineering Co., Ltd
Designer: Nie Jianping, Jiang Luchuan
Project Location: Hangzhou
Project Area: 840 m²
Main Materials: marble, oak, wall covering, wallpaper, texture paint, art glass

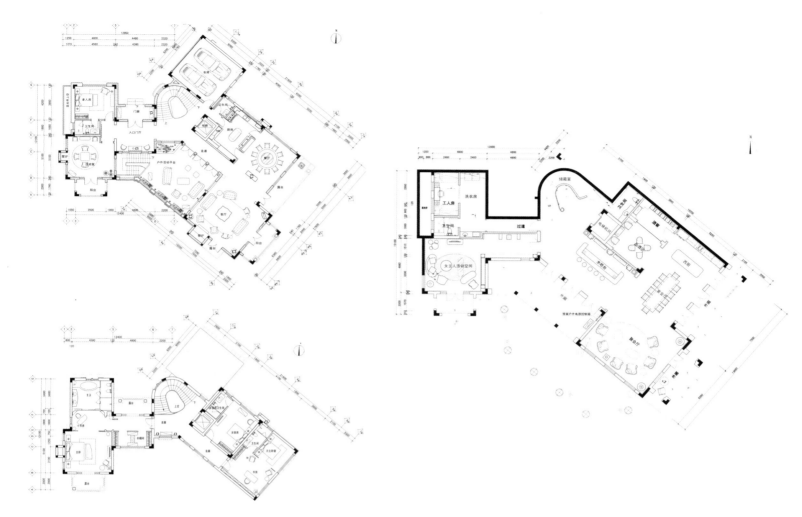

以奢为美 新古典豪宅设计

在人们的生活方式不断变化的今天，令人惊讶的是有一种真正的"法国式生活艺术"，虽历尽时代变迁，但却依然时尚。这种"法国式生活艺术"不仅以时装、香水、烹调、尖端科技为标志，而且亦反映在法国温馨优雅的生活风尚之中。这是一种令全世界都羡慕不已的生活艺术！它既传统典雅，又充满现代气息。它透过易于辨认的强烈特征，再现着某个辉煌的时代。它既追求舒适安逸，又注重沟通融洽，激发着人们享受生活的愿望。

在西野风韵，设计师希望将这种生活艺术传承下来。设计之初，设计师重新领略了卢浮宫、凡尔赛宫、枫丹白露宫等悠久的建筑艺术的风采，力求表现出声名卓著的法国精神，营造出一种既和谐大方又舒适典雅的贵族气氛。

设计师借鉴了被称为理性美代表的凡尔赛宫的艺术特点——整齐的轮廓造型，庄重雄伟；没有巴洛克及洛可可繁复的雕饰，以直线与圆角作为基本元素。强调材质的温润与造型的优美，彰显出华贵的气质与人文底蕴。

"法国式生活艺术"的一个重要表现就是家具。家具是生活情境最重要的展示者，设计师选用了原装进口的法国家具品牌"Mis En Demeure"，没有冰冷的材质，没有尖锐的线条，而是从自然的质地和柔和的线条中营造出极具历史感的氛围。这些产品全都是由高素质的手工艺人制作而成的，为了实现仿古效果，每件产品都要经过五到六次的上色。同时，手工艺人们又通过自己无可复制的手工技艺，在每件作品中强化了个人风格，让每件产品的背后都有一个故事。木材主要选用了橡木、松木和山毛榉木，以及法国进口的高档木材，如桃花芯木、红木和紫罗兰木等。

Nowadays life styles are changing almost everyday, it is surprising that there is a real "French art of living", while going through changing times, it is still fashionable. This kind of "French art of living" is not only symbolized in fashion, perfume, cooking, cutting-edge technology, but also reflected in the warm and elegant French lifestyle. This is envied by the world to have art of living! Both traditional and elegant, it is also full of modern flavor. It is easily recognizable by the strong features which are the reproduction of a glorious era. Demanding both comfortable and cozy, and also focusing on communication and rapport, the style stimulates people's desire to enjoy life.

In the Wild West Charm, the designers hope for the inherit of the living art. At the beginning of the design, the designers re-appreciate the Louvre, Versailles, Fontainebleau Palace and other ancient architectural art style, and strived to demonstrate the spirit of the prestigious French, to create a harmonious and generous, comfortable and elegant aristocratic atmosphere.

The designers learn from what is known as representatives of rational beauty - the art of Versailles - neat contour shape, solemn with majesty; without Baroque and Rococo complex carving, it just adopts straight lines and rounds as basic elements. It emphasizes on the beautiful shape of materials, highlights the luxury temperament and cultural heritage.

One of the most important elements of "French art of living" is the manifestation of furniture. Furniture is the most powerful display of life, so the designers used the imported French furniture brand "Mis En Demeure", which is no old material or sharp lines, but to create a highly sense of history and atmosphere through the natural texture and soft lines. These products were all made by highly qualified craftsmen. In order to make antique effect, each product must go through color process for five to six times. At the same time, the craftsmen enhanced personal style through its own copy of craftsmanship in every piece of work, so that each product behind itself has a story. Major timbers were selected such as oak, pine and beech wood, and first-class wood imported from France, such as mahogany, rosewood and violet wood and so on.

BUCOLIC LIFE IN SOUTH AMERICA

南美的乡村生活

项目名称：保利阆峰云墅L18示范单位
设计单位：广州市韦格斯杨设计有限公司
项目地点：长沙
项目面积：450 m²
撰　　文：陈正茂
主要材料：米黄大理石、石材马赛克、板岩、涂料、木饰面、定制木地板、铁艺、墙纸

Project Name: Poly Langyunfeng L18 Showflat
Design Company: GrandGhostCanyon Designers Associates Ltd.
Project Location: Changsha
Project Area: 450 m²
Writer: Chen Zhengmao
Main Materials: beige marble, stone mosaic, slate, paint, wooden finishes, customized wooden floor, iron, wallpaper

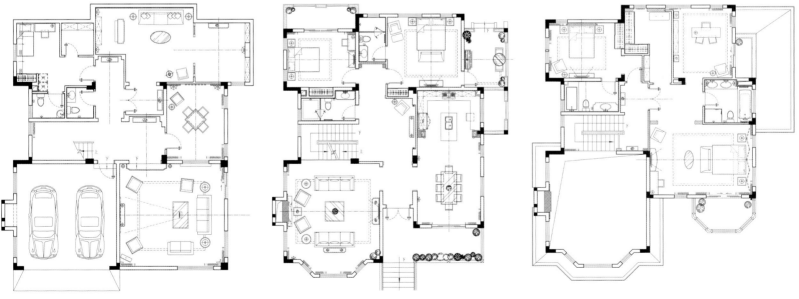

以奢为美 新古典豪宅设计

该案以深木色与石材的天然纹理以及墙纸的肌理相互搭配，形成了一种贴近南美乡村生活的空间氛围。

客厅是挑空大厅，设计师运用了巧妙的设计手法来展现顶棚的结构美，很好地避免了原建筑客厅的狭长感。地面运用了石材及经过特殊工艺处理的拼花木地板，与墙面苍劲有力的大面积的文化石材形成了强烈的对比，使其材质的运用多元化，而又具有整体感。阳光穿过高大的落地窗，照射在皮沙发上、墙上的动物标本上以及充满生机的绿色植物上，远离都市的休闲乡村生活就是这样的惬意与从容。

餐厅顶棚以木材为主要材料，其独有的天然纹理使人感觉与大自然相贴近。地面运用了仿古砖及光面理石拼接，丰富的拼花造型使空间活力四射。墙身选用高档花纹墙纸，与古典风格的餐桌及餐椅相搭配，营造出了温馨的用餐情调。

楼梯空间的墙身采用了肌理漆，与扶手的木材相对比，铁艺栏杆的特殊造型在灯光的映衬下仿佛形成了上下跳动的音符，为空间增添了不少活力气氛。

主人房的顶棚以木材为主材质，经过设计师的巧妙构思，原建筑的缺角全被隐藏在富有结构美的顶棚里面。地面采用木地板，让睡眠空间更加安静。墙身选用经典的西式墙纸，在硕大的床及贵妃椅的衬托下更显高贵。

This project weaves dark wood and the natural stone color with wallpapers, giving an impression of bucolic life in South America.

The roof of living room is much higher than elsewhere, which renders designers a chance to accentuate the structure of ceiling and makes the living room wider. The floor, which is made of stone and special-treated flowery wooden board, contrasts with large-scale wall made of natural stone. This project presents the diverse elements of materials, and the harmony between each of them. The sunshine traveling through the French window pours on the thick leather sofa, specimen on the wall and green plants, which is the cozy life you want to live.

The ceiling of the dining room is mainly made of wooden materials. The natural texture of the wood makes people feel closer to the nature. The floor is made of replica of old bricks and polished marble, which enliven the whole dining room. High-quality wallpaper is used to decorate the wall, and it works well with the ancient style dining table and chairs to create a sweet feeling for dining.

The walls beside the stairs are made of texture painting, compared with the wooden material of the handrail, the special design reflected by the light, iron banisters make itself be like lively music notes in the space as well, adding vitality to the space.

The ceiling of the bedroom is made of wood, after many thoughts, the original corners are covered by well-structured new ceiling. The wooden floor sooth your sleep. Wallpapers are of western style which makes the space look more elegant against the large bed and luxury couch.

WORLD OF CLEAR WATER AND LAKE HOUSE

一湖碧水 一宅天下

项目名称：维多利亚港湾别墅样板房
设计单位：HBS宁波红宝石装饰设计有限公司
设 计 师：张向东
项目地点：芜湖
项目面积：550 m²
主要材料：大理石、墙纸、复古砖、实木地板

Project Name: Victoria Harbour Villa Model House
Design Company: HBS + Ningbo Apyroti Decoration and Design Co., Ltd.
Designer: Zhang Xiangdong
Project Location: Wuhu
Project Area: 550 m²
Main Materials: marble, wallpaper, retro tile, wood flooring

维多利亚港湾别墅位于芜湖，该项目拥有三面环湖的28栋独栋豪宅，外观雍容，内部空间分布合理。坡顶、露台营造出别致的层次感，在规划布局上采用自由式，通过建筑单体的弯折、转向与错接，楼型高低错落等因素，营造出浓郁的欧洲之风。建筑细节的设计充分尊重自然、拥抱健康，低调中传达出浓浓的浪漫之风。

样板房在设计上不但要考虑建筑内与外的结合，更要考虑项目本身的高贵品质，体现出该案营造的ART DECO风格的大气、尊贵、宁静的特色。地下室被打造成融视听、健身、运动于一体的休闲空间，体现出人们追求健康、休闲的生活方式。挑高客厅的设计，使阳光爱上室内的每一寸空间。结合精美的工艺品及温暖的色调，营造出迷人的、充满艺术气质的氛围。

Victoria Harbour villa is located in Wuhu, and the project has a lake surrounded by 28 single-flat mansions, whose appearance is gorgeous with reasonable internal spatial distribution. The top of the hill and the terrace create a chic layering. The use of freestyle in the planning layout through the construction of the monomer bending, turning and wrongly connection, the scattered and unbalanced levels are in typical European style. Architectural details of the design are with full respect for nature, healthy life style, and a deep sense of romance with low profile.

When design Model houses we not only take consideration for the combination of inside and outside buildings, but also the noble quality of project itself, which can reflect the ART DECO style with grand, distinguished, and quiet characters in this project. Basement is designed as a place for leisure time: film, fitness, and sports, which reflect the modern pursuit for healthy and leisure lifestyle. The high-ceilinged design for living room makes the sunshine get touch of every inch of interior space. The combination of fine arts and crafts and warm colors create charming and artistic atmosphere.

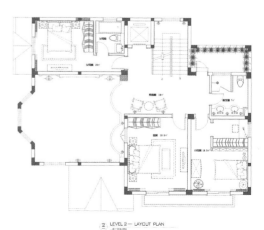

以奢为美 新古典豪宅设计

FRENCH ART OF THE MODERN INTERPRETATION OF THE COURT

法式宫廷艺术的现代演绎

项目名称：上海奉贤法式别墅

设计单位：巫小伟/威利斯创意设计中心

项目地点：上海

户　　型：独立别墅

项目面积：500 m²

主要材料：罗马柱、大理石、壁纸、防水石膏板、饰面板、软包、金箔贴片、水晶灯、铁艺楼梯

Project Name: Shanghai Fengxian French-Style Villa
Design Company: Wu Xiaowei/Venice Creative and Design Center
Project Location: Shanghai
Unit Type: Independent Villa
Project Area: 500 m²
Main Materials: Roman column, marble, wallpaper, waterproof plasterboard, panels, soft bag, the gold foil patch, crystal lamp, wrought iron stairway

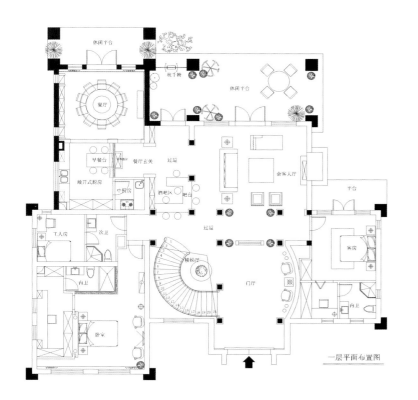

一层平面布置图

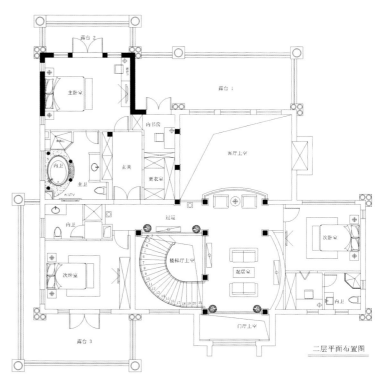

二层平面布置图

该案为上海奉贤的一套古典欧式独栋别墅,室外小桥流水,鸟语花香,室内极尽奢华,雍容高贵。古典欧式风格的大气浪漫在此展现得淋漓尽致,每一个细节都精心雕琢,犹如置身于欧洲的古典宫廷,给人以不尽的视觉享受和感叹。

推门而入,门厅即把整个大宅的气度展现得气象万千,穹顶灯池、罗马柱、大理石地面、旋转楼梯、空调排风扇等都经过精心的雕花处理,还有华美典雅的座椅、水晶灯、精致的饰品,小小的门厅不遗余力地诉说着豪宅的气度,奢华气息与艺术享受水乳交融。

门厅与客厅的过渡区间简约而自然,顶部采用石膏线吊顶,加上金箔纸贴片、水晶灯、筒灯以及隐藏的灯带,精致的灯光组合营造出理想的生活空间。客厅的挑高空间使视觉上极为开阔,布置却甚为紧凑。设计师采用了大量的横平竖直的线条来表现空间的张力,壁炉上方采用明镜装饰,延伸了视觉空间。顶部的大型水晶吊灯则有效地填补了顶部空间,欧式古典的布艺沙发、色彩浓艳的窗帷、台灯、绿色植物等的布置都是精心挑选的。

客厅和餐厅之间在原有结构的基础上设置了吧台,很好地将不同的功能空间进行了过渡。餐厅同样极尽奢华,弧形灯池、装饰酒柜、弧形落地窗等勾画出极佳的用餐氛围。

铁艺楼梯旋转而上,二楼私密区的设计则更为温馨,休闲区、主卧、主卫、儿童房等,无论是在设计上还是后期的软装配饰,均流淌着贵族的气质。

The project is a classical European-style located in Fengxian of Shanghai. There is a small bridge over the flowing stream, lots of flowers and birds outside. The inside of the house is graceful and noble. Every corner of the house is so carefully crafted as if you were in a European classical palace.

When you get in the house, you can see the Roman column, the marble floor, the spiral stair and the air fan. All the things in the hall are the miniature of this elegant villa. There are also elegant seats, crystal lamps and beautiful decorations which reflect the connection of grace and art.

Hallway connects the living room naturally, and its top is plaster ceiling with decorations of gold foil patch, crystal lamp and invisible lamp group. The extension in height of living room makes the space look wider. Designer use lots of straight lines to show the tension of space, and the place of a silver mirror above the fireplace stretches the field of vision. Large-scale crystal lamp fills the blank ceiling. Ancient European fabric sofa, colourful curtain, table lamp, plants are all carefully chosen.

A bar counter is laid between the drawing room and dining room as a transition of different functions. The dining room is also graceful for decoration wine ark, arc French window.

Wrought iron stair goes to the second floor where the main bedroom, the main bathroom, the children's room are also egelant.

AMERICAN NEO-CLASSICISM MAKING GREAT LIFE

美式新古典演绎精彩人生

项目名称：深圳中信红树湾住宅
设计单位：深圳市朗昇空间艺术设计有限公司
设 计 师：袁静、钟建福
项目面积：180 m^2
主要材料：大理石、墙纸、涂料等

Project Name: Zhongxin Hongshuwan Home of Shenzhen
Design Company: Shenzhen Lonson Environmental Art Design Co.,Ltd
Designeners: Yuan Jing, Zhong Jianfu
Project Area: 180 m^2
Main Materials: marble, wallpaper, painting, etc.

美国人喜好追求一种自由而休闲、随意而舒适的生活方式，美式风格汇集了法式、英式、意式等不同风格的精华，浸润了美国文化的艺术特点。其优美的线条、厚重的色调、粗犷舒适的家私，演绎出悠闲而舒适的生活情调。

该案是一套新古典美式风格的室内设计，位于深圳中信红树湾，面积180平方米。为使客厅看起来更宽阔些，设计师将阳台改造成开放式书房，既与客厅联为一体，又可独立使用。客厅墙面及天花以白色与米黄色来搭配，使空间整体看起来明亮、大方、层次丰富，整个空间给人以开放、宽容的气度。

卧室房间多使用不同图案的墙纸，使不同空间充满不同的情趣，以营造出不同的空间氛围。主卧室中，显得高雅而和谐，儿童房中，则显得天真而浪漫。

美式风格的家私一般均为实木制作，线条优美大气，体积宽大气派、温馨舒适。金属灯具、金属风扇、玻璃制品以及中式画作，均表现出美式风格兼容并蓄的文化特性。

美式风格，演绎的是一种居住态度，更是一种精彩的生活方式。

American loves the unrestrained and comfortable life. Designer merges the essence of French, British, Italian style with American native culture together. The beautiful lines, thick paint, comfortable furniture make a wonderful living environment.

This case is an American Neo-classicism design, located in Shenzhen Zhongxin Hongshuwan district, which covers 180 square meters. The designers remake the balcony into an open study room in order to widen the living room. White wall matches creamy ceiling well, lightening the whole room.

The designers use beautiful wallpaper with different patterns to create different pleasures in every unique space. The main bedroom is full of elegance while children's room is full of innocence and pureness.

American furniture is usually made of solid wood, and its lines are exquisite and comfortable. Metal lamps, metal fans, glassworks and Chinese paintings reflect American style's compatibility of other styles.

American style shows an attitude towards life and creates a living way.

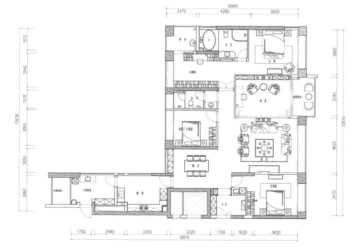

以奢为美 新古典豪宅设计

SIMPLE AND ELEGANT AMERICAN STYLE

古朴拙雅之美式风

项目名称：云宙天朗高尔夫别墅
设计公司：福州合诚环境艺术装饰有限公司
设 计 师：王伟
项目地点：福州
项目面积：300 m²
主要材料：意大利蜜蜂瓷砖、圣象实木地板、大风范家私
摄 影 师：施凯、李玲玉

Project Name: Yunmu Tianlang Golf Villa
Design Company: Fuzhou Hecheng Environmental Art Decoration Co., Ltd.
Designer: Wang Wei
Project Location: Fuzhou
Project Area: 300 m²
Main Materials: Italian bees tile, Power Dekor wood floors, large style furniture
Photographer: Shi Kai, Li Lingyu

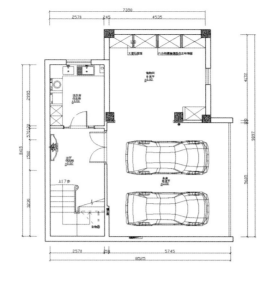
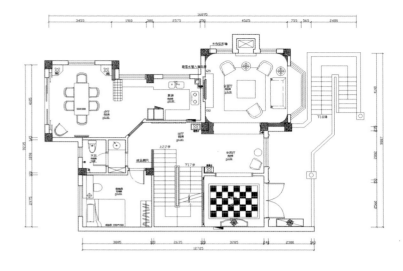
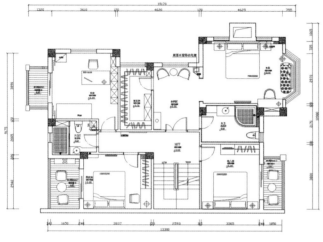
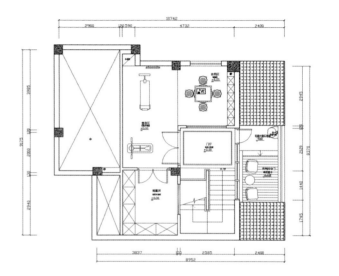

这里没有璀璨耀眼的水晶灯,也没有奢华的大理石背景墙,但这里有建筑中内与外的融汇,这里有褪去奢华、质朴拙雅的原味,这里有心对家的眷顾……

设计师通过改造原建筑的不足,重新整合空间的布局。当入户玄关的仿古砖,偶遇铜制挂画,使空间的搭配变得充满趣味。客厅的家具陈设以古朴拙雅的美式风格为主,粗犷质朴的乡村壁炉、温暖宽适的布艺沙发,以及细腻柔美的提花地毯、暗花窗帘……眼前的一切让人无法怀疑混搭的美。

餐厅与会客区不在同一空间内,因此设计师大胆地打开厨房与餐厅间的界面,并利用拱门延展了空间,在三面环窗的餐厅里,阳光跳跃在空间中的每个角落,在这里身心回归自然不再是种奢望。

通往二楼是道狭长的梯间,墙身巧妙地开设了两个小拱洞,既与周边拱门相呼应,又弱化了过于狭长的楼道。二楼主人卧室中,设计师更多地从使用功能的角度合理安排家私配置及软装陈设,无需过多的雕琢,空间的美已跃然于上。

No bright shining crystal light, no luxury marble background wall, but the fusion of inside and outside building is here; the rustic flavor with complicated taste without luxury is here; the determination for the tender care of home is here...

The designer modified the original construction deficiencies and reconstructed the spatial layout. The match of the antique brick and the copper paintings at the entrance make the design full of fun. The main furnishings in living room are in humble elegant American style - rural rustic fireplace, warm and wide fabric sofa, and the delicate soft jacquard carpet, floral curtains...all of which make the beauty of mix and match become evident.

Dining room and guest reception area are not in the same place, so the designer just opened the interface between the kitchen and dining room, and used arch to stretch the space. Surrounded by three windows, sunshine can touch every inch of space in the dining room thus it's not just a dream any more to enjoy coming back to nature both physically and mentally.

A narrow staircase leads to the second floor, and two small arches are cleverly set up on the walls, which echoes to the arches surrounded and also weaken the long corridor. In the master bedroom on the second floor, the designer takes more consideration on the functionalities to make reasonable arrangement for furniture and soft furnishings - without too much polish, the beauty is just ready for appreciation.

以奢为美 | 新古典豪宅设计

151

ACROSS THE TRADITIONAL AND THE MODERN NEO-CLASSICAL STYLE

跨越传统与时尚的现代新古典

项目名称：福州（融侨城）
设计公司：品川设计顾问有限公司
设 计 师：周华美
项目地点：福州
项目面积：280 m²
主要材料：银箔、镜面、铁艺、紫檀木面、皮革、墙纸、大理石

Project Name: Fuzhou (Rongqiao City)
Design Company: Pure Charm Design
Designer: Zhou Huamei
Project Location: Fuzhou
Project Area: 280 m²
Main Materials: silver foil, mirror, iron, rosewood veneer, leather, wallpaper, marble

一层平面布置图

二层平面布置图

以奢为美 | 新古典豪宅设计

该案在满足空间使用功能的同时,将客厅挑空,为之后的风格装饰进行铺垫,使整体上感觉浑厚而大气。通透的采光效果让空间富有生命力。设计师将铁艺的优美曲线恰如其分地贯穿于整套别墅的设计中,这种镂空的艺术品巧妙地分割了一楼客厅与餐厅的功能区,在光影的映射下,这种通透式的隔断使空间产生了融会贯通的曼妙感。

色调的变化犹如空间中跳跃的音符,设计师利用柔和的米黄色调和局部空间中沉稳庄重的黑色产生的色彩对比,透过富有张力的线面搭配和柔美的软面装饰,体现了理性与感性兼具的特质。强烈的空间层次关系,铺排上暖黄的灯光,勾勒出了典雅低调的氛围。

在设计上,无论是家具还是配饰,均以其优雅、唯美的姿态,平和而富有内涵的气韵,描绘出居室主人高雅的身份。另外,设计师选用了不同种类的材料,在不同质感的碰撞中激起创意的火花,将多元性和趣味性静静地融入于家居的温馨气氛中。在低调的奢华中渗透着与众不同的品位与格调。深色的木面银箔烘托出一种高贵的氛围,通过银镜的反射性来拓展视觉的张力,使整个空间给人以开放宽容的非凡气度。

On the one hand, the design meets the use of space capabilities; on the other hand the ceiling of the living room is very high as to pave the way for the style of decoration, so the overall design is grand and open. The lighting effects make the department quite transparent and full of vitality. The graceful curves of the wrought iron are appropriately used throughout the entire design of the villa, and this kind of hollow art delicately split the living room on the first floor and the dining area, and under the shadow of the partitions an amazing transparent sense is created.

The change of colors is like the jumpy notes in space. The designer adopts the soft beige to balance the contrast of the partial black color with quiet dignity. By the match of lines and surfaces, and soft surface decoration reflects the characteristics with both rational and emotional qualities. The obvious spatial layout on the warm yellow light sketches out an elegant low-key atmosphere.

In the design, both the furniture and accessories presents the identity of the master bedroom through their elegant, aesthetic attitude, calming and meaningful styles. In addition, the designer uses different types of materials to make different textures so that the diversity and fun are embedded into the warm home. The low-key luxury style also displays distinctive taste. The dark wooden surface with silver foil expresses a kind of noble style, through the reflection of silver mirror to expand the visual tension, making the whole space present an open, wide and extraordinary disposition.

TIME FLIES

岁月如歌

项目名称：福州某别墅
设计公司：福州华悦空间艺术机构
设 计 师 ：胡建国
项目地点：福州
建筑面积：300 m²
主要材料：实木地板、壁纸、大理石

Project Name: A Villa in Fuzhou
Design Company: Art Institute of Fuzhou Hua Yue
Designer: Hu Jianguo
Project Location: Fuzhou
Building Area: 300 m²
Main Materials: solid wood flooring, wallpaper, marble

一层平面布置图 SCALE 1:100

二层平面布置图 SCALE 1:100

一层吊顶布置图 SCALE 1:100

二层吊顶布置图 SCALE 1:100

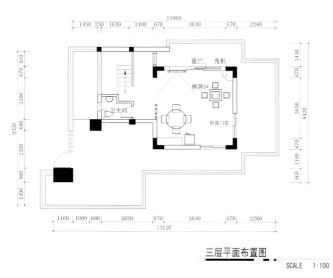

三层平面布置图　SCALE 1:100

三层吊顶布置图　SCALE 1:100

悠长的岁月可以带走很多，却也留下了许多值得回忆的东西……该案为低调复古简欧风格，追求在时尚生活下，那呼之欲出的浓郁的欧陆风情，在优雅中感受音符的律动，在平静中演绎岁月的痕迹。

设计师在做这个方案时考虑了很多，希望这个作品既不流于表面的奢华，也不追求形式感的复古，而是一个深具典雅美学底蕴、带有浓郁历史文化痕迹的作品。不仅要有良好的视觉效果，更重要的是要体现出和谐浪漫的心理感受。设计师充分结合了建筑自身的空间特色，充分满足实用功能，营造出比例舒适、色调和谐的优雅空间。

Time can take away a lot, but it also left many memorable things...The case is designed in low-key retro simple European style to reflect the pursuit of the rich European style, to feel the rhythm in the elegant atmosphere to interpret the traces of time in quiet environment.

The designer took a lot into consideration about this project to make it neither superficial luxury, nor looks in retro form, but a masterpiece with richly elegant aesthetics, historical and cultural trails. The designer wanted to make it have good visual effects, more importantly, to reflect the psychological feelings of romantic harmony. The designer fully integrated space features of the building and met the practical function, thus created a apartment with comfortable ratio, elegant and harmonious color.

THE GLAMOUR OF LIVING IN MANOR

庄园生活之魅

项目名称：比华利别墅
设计单位：BMD design
设 计 师：鲍明达
项目面积：800 m²
主要材料：仿古砖、拼花地板、墙纸

Project Name: Beverly Villa
Design Company: BMD design
Designer: Bao Mingda
Project Area: 800 m²
Main Materials: antique brick, block floor, wallpaper

地下室平面布置图

一层平面布置图

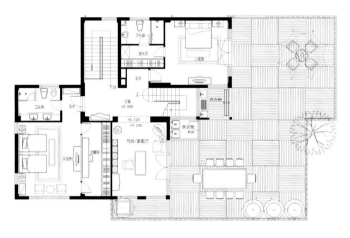
二层平面布置图

三层平面布置图

名流汇聚的比华利（Beverly）早已成为Rich&Famous的代名词，在宁波，比华利别墅区无疑是最好的居住区之一，业主对于生活品质和装饰设计的要求自然也毋庸置疑。

该案中，设计师将大气浑厚的北美风情与现代人居理念完美融合，从外部的景观改造到出入口的设计，从天花的线条刻画到地面仿古砖及拼花地板的应用，配合精细的手织地毯、唯美的陈设以及大气的美式家具，体现出名流生活的奢华与闲适感。

别墅内外通过大面积的落地玻璃巧妙地将室外景色引入室内，建筑、园林及周边原始植被的景观变化无穷，使业主在享受静谧休憩空间的同时，也拥有极佳的景观视野。生活于此，临水而居，品花而坐，使居住者真正体会到了"坐览水云闲，行闻风露雅"的优美意境，充分诠释出庄园生活的闲适与淡泊。

Beverly with various kinds of celebrities has been the symbol of Rich&Famous. In Ningbo, it is no doubt that Beverly Villa District is one of the best residential communities, in which the owner's demand for living quality and decorating design is undoubtedly high.

In this project, the designer connects the gorgeous North-American style with the modern residential concept, from the angle of the reconstruction design of outside landscape to the design of passage way, from the carving of ceiling line to the use of antique brick and block floor. The application of fine hand woven carpet, aesthetic finishing and American-style furniture show the luxurious living and comfortable feelings.

The villa introduces the outside view skillfully through big and wide French window. The change of landscape in architecture, gardens and original vegetations make the owner not only get a quiet and comfortable space, but get a good view. Living in this villa which fully shows comfort and satisfaction of living, you are surrounded by water and flowers and get the feeling of grace and elegance.

THE NEW EUROPEAN-STYLE STRIKES

新欧式风格来袭

项目名称：长乐世纪公馆某住宅
设计单位：福建国广一叶建筑装饰设计工程有限公司
设 计 师：唐垄烽
方案审定：叶斌
项目地点：福州
项目面积：300 m²
主要材料：爵士白大理石、镜面玻璃、意大利艺术砖、钨钢不锈钢、艺术墙纸等

Project Name: An Apartment of Changle Century Mansion
Design Company: Fujian Guoguang Yiye Decoration Design Engineering Co., Ltd.
Designer: Tang Longfeng
Program Validation: Ye Bin
Project Location: Fuzhou
Project Area: 300 m²
Main Materials: Jazz White marble, mirror glass, Italian art tiles, tungsten stainless steel, tungsten, art wallpaper, etc.

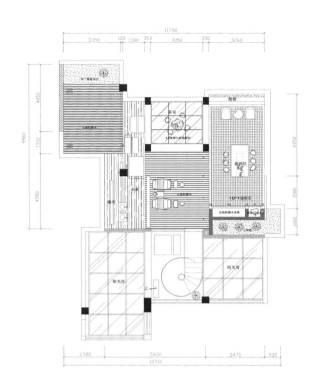

新古典豪宅设计

室内空间就犹如凝固的音乐,它散发出的力和美就像音乐的节奏和韵律,冲击着人们的心灵,带给人们美的感受。

该案设计师采用新欧式的处理手法来演绎整体空间。新欧式设计风格以欧洲丰富的艺术文化为底蕴,根据地域文化的不同又有不同的表现形式。在该案中,设计师用现代的手法和材质还原古典气质,整个空间的设计具备了古典和现代的双重审美效果,这种效果与现代人所追求的品质与休闲生活的完美结合相适应,使业主在享受物质生活的同时,也得到了精神上的慰藉。

Interior space is like frozen music, the power and beauty it displays is like rhythm and pace of the music, which impacts people's minds and gives people beautiful experience.

The designer adopts new European approach to interpret the overall style. New European design style is based on the rich artistic culture of the European heritage according to the different regions with different cultural manifestations. In this case, the designer restored the classical temperament with modern techniques and materials, so the entire design has the classical and modern aesthetic effects, which meets the pursuit of combination of leisure of quality of modern life, so that the owner can enjoy material life and get spiritual comfort at the same time.

WEST MELODY IN MODERN HOME

现代府邸里的西洋调

项目名称：玛斯兰德郭先生雅宅
设计单位：一格建筑规划设计有限公司
设 计 师：廖昕耀

Project Name: Masterland Guo's Ornamental House
Design Company: Yige Planning and Designing Corp
Designer: Liao Xinyao

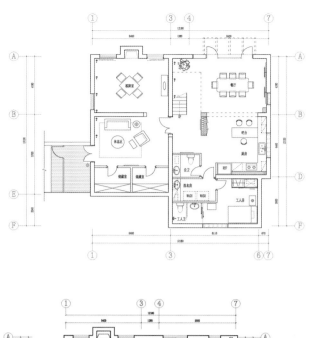
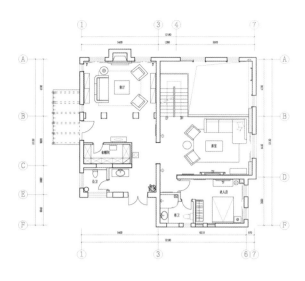
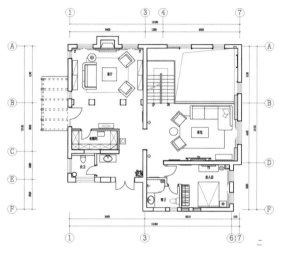
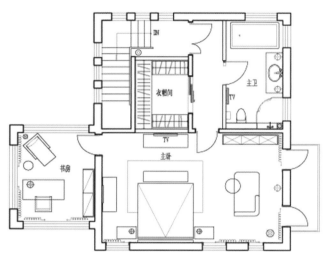

该案中，清晰的轮廓与淡雅的色彩相搭配，细致的米黄色硅藻泥墙面，体现出时尚和气派。整个空间没有披金戴银的奢华，却有着不张扬的舒适感，依稀可见豪门的尊贵。鲜明的直线条设计彰显出大气的生活态度，生活空间中到处弥漫着舒适、轻松的贵族气息。

设计师对这套别墅风格的拿捏，体现最明显的是对有鲜明设计风格的配饰的运用上。每一件家具的雕花都精致典雅，但样式绝不雷同；各式各样的蜡台为整个空间增添了几许浪漫的情调；随处可见的定制的装饰画，彰显出业主不俗的欣赏品位。从米黄色墙面、软包沙发、地毯到天花板上的欧式线条及水晶吊灯，整个空间都弥漫着一种贵气的氛围。

厨房与餐厅空间中的落地窗与米黄色的顶梁造型以及浅色墙面，使空间显得开敞而明亮，极富层次感。宽大的落地窗，使室内外景观相呼应，打造出现代人追求的人与自然和谐共处的环境。

主人卧房采用精致的复古设计，同时又散发着现代气息。通过材料的应用，使古典的元素和语汇与现代的家居空间完美地结合一起。古朴的暗红色地砖和精致的家具，使整个空间温暖和煦，墙面上简洁的壁炉造型为空间平添了几许生活的气息。

In this project, distinct lines go well with elegant color; fashion and dignity are obviously reflected by creamy yellow diatom ooze wall. You can experience the low-key comfort without redundant decorations and feel dignity of the noble. Distinctive lines manifest generous attitude towards life and the whole living space is full of aristocratic air.

The key of handling this house is the application of decorations that have outstanding characteristics. Every furniture is carved with delicate patterns which are unique and different from each other. Various candlesticks add wonderful romance to this room; customized decorative pictures reflect the elegant taste of the owner. The whole room is full of aristocratic air that you can feel from the creamy wall, soft sofa and carpet to European ceiling and crystal chandelier.

French window, creamy top beam and light-colored wall between kitchen and dining room lighten the whole space and make it hierarchical. Capacious window gives a good view to appreciate the beautiful scenery outside, creating an environment which modern people always pursuit to approach the nature.

Host bedroom is designed in a vintage style without abandoning the contemporary soul. The designer takes measures to combine the ancient elements and modern living space together. The elegant dark red floor and delicate furniture warm up the whole room, and the fireplace adds more living pleasure to this room.

以奢为美 新古典豪宅设计

MODERN INTERPRETATION OF MEDIEVAL CASTLE

中世纪古堡的现代演绎

项目名称：皇冠钻石双层建元地产
设计单位：一格建筑规划设计有限公司
设 计 师 ：廖昕耀
项目面积：300 m²

Project Name: Double Diamond Crown Jianyuan Estate
Design Company: Yige Planning & Designing Company
Designer: Liu Xinyao
Project Area: 300 m²

一层平面布置图 1:80

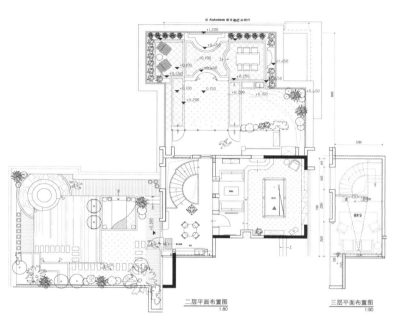

二层平面布置图 1:80　　三层平面布置图 1:80

以奢为美 新古典豪宅设计

该案的设计以中世纪古堡的空间概念，结合现代的生活方式，抽离过多的繁复雕饰，空间层次逐级递进，以华美繁复的古典元素装饰大气简洁的空间，浓烈的色彩呈现出典雅与高贵感。

客厅空间从整体到细节都经过设计师的悉心调整，以米色为主色调，细致、华丽的大理石、细腻的香槟银铂是基本的装饰元素，雅致的摆设散发出别具慧眼的艺术修养。整个空间没有披金带银的奢华，却有着不张扬的舒适感。每一处细节，如同绽放在春风中的花朵，缓慢地舒展，每一处都透露着新鲜，且每一处都是那么的熟悉，演绎出一幅和谐的家庭画面。

设计师用现代的手法演绎西方文化，让久远的古典艺术于庄重中带有时代感和生活气息。深色的家具诠释了纯净无瑕的经典品位，色彩鲜艳而又不失庄重，天然纹理的大理石在这样的环境里，褪去了原始的味道，融入时尚的潮流。正如主人儒雅的性格，满溢着潇洒的气质。

The design of this apartment is combined space concept of Medieval castle design with modern lifestyle without too complicated carvings, but the designer uses classical beautiful complex decorative elements to perfect the succinct space, and the strong colors presents elegant and noble sense of feeling.

The living space has been adjusted from the whole to the design details which is mainly based on beige color with detailed and ornate marble, and delicate champagne silver platinum is taken as the basic decorative elements, the elegant furnishings exudes unique taste for artistic accomplishment. The entire apartment is designed without the luxury gold or silver but it is with great comfort. Every detail is just like the flowers blooming in the spring, which are slowly stretching and revealing fresh and familiar sense as a harmonious family picture.

The designer uses modern techniques to interpret Western culture, so the ancient classical art is decorated with contemporary sense and touch of life. The dark furniture shows the classic pristine taste. The colors are bright and dignified, and the marble with natural grain in such an environment lost its original flavor, but is embedded with some fashionable elements. The space is refined as the owner's personality - elegant as well as full of cool temperament.

GEOMETRIC SHADOW

几何光影

项目名称：温州某住宅
设计单位：印象设计
设 计 师：李孝都
项目面积：300 m²
项目地点：温州
主要材料：实木、不锈钢等
摄 影 师：阿龙

Project Name: A Residential Apartment in Wenzhou
Design Company: The Impression Design
Designer: Li Xiaodu
Project Area: 300 m²
Project Location: Wenzhou
Main Materials: wood, steel, etc.
Photographer: Aaron

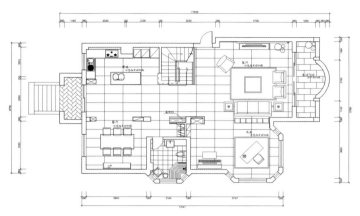
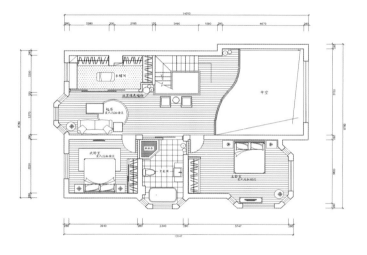

该案设计师以营造简约、时尚、庄重的空间氛围为设计目的，整体空间布局开阔而通畅。在色调上，以黑、白、咖啡为主色调，这三种色调搭配起来的空间形象沉稳而庄重，与居住者的年龄相符合，局部点缀的亮银色，又为该空间增添了时尚气息。

在软装的选用上，设计师精挑细选了一些富有情趣的装饰品，比如茶几上的装饰瓶，其柔和而优雅的曲线造型犹如一位翩翩起舞的少女。在灯光的处理上，暖色调的灯光打破了空间过于沉稳、肃静的冷漠感，使温馨的居室气氛得到呈现。

"家"是一个使疲惫的身心沉静下来的港湾，是一个抛却所有烦恼、卸下所有心防的温馨家园。

The apartment was designed to be succinct, stylish, and dignified. The overall layout of the space is open and unobstructed. The main tone is based on black, white, coffee colors, the combination of which is calm and solemn, and it's consistent with the age of residents. Bright silver as partial decoration also added stylish into the atmosphere. The designer delicately selected some interesting accessories for the soft decorations, such as decorative bottles on the coffee table, whose soft and elegant curved shape is like a dancing girl. For the lighting, warm light broke sense of apathy - over calm and quiet, so the atmosphere of the room has been presented.

"Home" is a harbor where the tired body and mind can calm down, forget all the trouble, and remove all guards for the warmth of home.

LOOKING FOR THE BEST TIME IN LIFE

寻访生命里最美的时光

项目名称：波托菲诺复式
设计单位：深圳朗昇环境艺术设计有限公司
设 计 师：袁静、钟建福
项目面积：300 m²
主要材料：大理石、地砖、玻璃等

Project Name: Portofino Duplex
Design Company: Shenzhen Lonson Environmental Art Design Co., Ltd
Designer: Yuan Jing, Zhong Jianfu
Project Area: 300 m²
Main Materials: marble, floor tile, glass, etc.

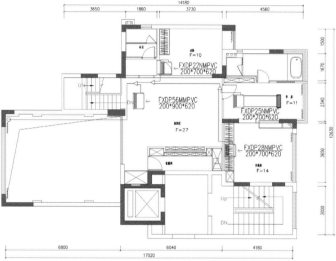

该案为三层复式空间，设计上以新古典奢华主义为主，表现出高贵复古、写意奢华的空间格调。它既有欧洲文化的艺术底蕴，又有新古典风格的开拓创新。设计师以传统古典作为本案的灵魂，以时尚作为外衣，把或中或西的元素糅合在一起，跳脱时空的界限，呈现出充满活力的姿态。通过对空间的细腻诠释，在复古风潮和实用理念的双重影响下，为追求品质生活的城市精英，缔造出了一个温馨、舒适、温馨的心灵港湾。

The case shows a three-storey double space characterized with neo-classical luxury style, which expresses an elegant, ancient, comfortable space. It involves not only European classical art but the creative innovation of neo-classical style. With fashionable appearance, the designer makes the traditional classic as the soul of the case, and connects the oriental elements with occidental elements, which is beyond the limits of space and time and shows vigorous atmosphere. Through the exquisite explanation of space and under the influence of the trend of retro-style and practical concept, the case provides a warm, comfortable and happy harbor in soul to the city elite who pursues the high quality of living.

ENJOYING ELEGANT PIANO MUSIC

畅享优雅的钢琴曲

项目名称：福州香榭丽舍
设计单位：楼语设计工作室
施工单位：朗雅别墅施工
设 计 师：林金华
项目面积：350 m²
主要材料：大理石、蜜蜂瓷砖、墙纸、造型灯
摄 影 师：施凯

Project Name: Fuzhou Champs Elysees
Design Company: Louyu Design Office
Construction Unit: Langya Villa Decoration
Designer: Lin Jinhua
Project Area: 350 m²
Main Materials: marble, Bee tile, wallpaper, modeling lamp
Photographer: Shi Kai

以奢为美 新古典豪宅设计

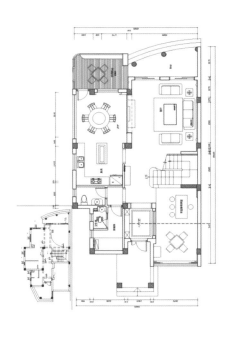
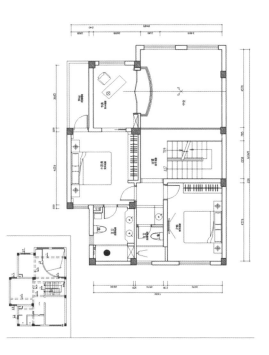
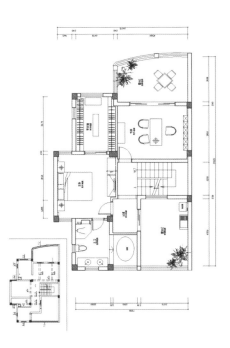

以奢为美 新古典豪宅设计

217

走进这个三层的别墅空间，简约的意境，弥漫着唯美的幸福。入门处一幅宽大的挂画，泛着浓浓的艺术气息，婉约不尽，又酷劲十足。一把欧式韵味的椅子，似乎在静静地等待主人的到来，方便而实用。休闲区中几张红色的实木矮凳，以及极富意境的壁画，使空间变得悠闲而静谧，心情紧张之余不妨到此休息一番，会别有一番滋味在心头。

设计师考虑到空间的限制，将楼梯进行了改变，并把餐厅进行外扩，让整体空间显得更加得宽敞。靠在敞亮的窗边，听着优雅的钢琴曲，看夕阳点点的金黄在屋子里静静地流淌，听着一个个音符悄然滑落，这是坠入心底的声音。这样静谧的气氛，让心情也不免沉静下来，畅享着居家的幸福。

径直走入，便来到客厅，特意挑高的空间，一盏水晶吊灯悬在空中，诉说着空间的优雅与尊贵。欧式风格具有很强的文化感与历史内涵。餐厅内的家具也有意识地选择了外形精致的矮凳和深红的圆形餐桌，配合别针式样的乳白色吊灯，使空间显得额外素雅。红木的整体厨柜、L形的柜台、白色的柜面，使烹饪空间洁净而便捷，既实用又使得空间的色彩极具美感。

主卫空间中浪漫的欧式情调十足，浴室的蓝色主色调，使人心情瞬间放松，空间显得非常舒适。客卫采用了绿色为主的色调，同样是让人心情愉悦的色彩，将居家休闲和舒适的氛围展现得淋漓尽致。

Entering the three-storey villa, the simple artistic conception conveys aesthetic happiness. On the entrance, a picture in wide shape gives out strong artistic taste, it is graceful enough and cool enough. A chair in European style is practical and convenient to use, it seems to wait for its owner quietly. In the recreation area there are several red stools made of solid timber, together with the mural which is rich in poetic imagery, the space becomes relaxing and quiet. Having a rest here can help people get rid of tension and get into another mood.

In consideration of the limited area, the designer made some change in the staircase and extended the dining room, which make the whole space wider. Leaning on the window in the spacious and bright space, you can listen to the elegant piano music, look at the sunset glow flowing quietly in the room. A series of notes dropping quietly, it is the sound that reaches people's inner heart, it makes people calm down and enjoy the happiness of home.

Walking straightly you can see the parlor, the high space is deliberately designed, a crystal droplamp suspending in the air conveys elegance and royalty of the space. The European style expresses strong cultural sense and historical connotation. Furniture in the dining room including exquisite chairs and crimson round table are intentionally chosen, matching with milky white droplamp, the space looks simple and elegant. Integral ambry made of redwood, counter in the shape of a capital "L" and white counter covering make the kitchen clean and convenient, practical and beautiful.

The main bathroom has strong European romantic style, taking blue as the keynote, people would feel relaxed and comfortable in it. Green is the main color in another bathroom, it is also a color which can give people pleasant feeling, leisure and comfort of home living is expressed thoroughly.

MODERN MINDS ON VILLA

豪宅新主张

项目名称：翠湖绿洲花园示范单位
设计单位：广州市韦格斯杨设计有限公司
项目地点：佛山
项目面积：309 m²
主要材料：埃及米黄石、黑金木纹、路易斯金、浅啡网、白木纹、阿富汗金、皮革、玫瑰金镜钢、墙纸、地毯

Project Name: Cuihu Oasis Gardon Showflat
Design Company: GrandGhostCanyon Designers Associates Ltd.
Project Location: Foshan
Project Area: 309 m²
Main Materials: Egypt perlato, Niello wood grain, Lewis's metal, Light emprador, Perlibino bianco, Afghanistan gold, leather, Rose gold mirrored steel, wallpaper, carpet

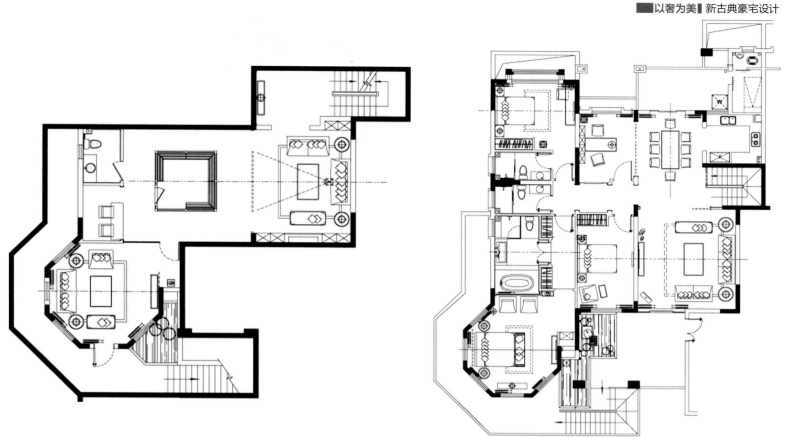

该样板间是为数不多的三层复式户型。负二层为车库,负一层为具有休闲娱乐功能的空间,首层由客厅、餐厅、卧室组成。在平面布置上,设计师在优化原先建筑平面功能布局的同时,又具有展示的效果。

该案中,设计师通过提取西式风格的经典元素,重新演绎了现代居室的豪华时尚感。运用现代的设计手法,在材料的选取上主要以石材和木饰面为主,着重强调空间的面与体量比例上的关系,呈现出稳重尊贵,又不失时尚的感觉,讲究气派、气度与细节。空间中高雅的米黄色调,天花上低垂下来的水晶吊灯,墙面上的细节处理,地面上的拼花图案,都体现出了本居室的不凡气度并透露出现代人所追求的生活质量。力求每一处细节都能诠释出高雅与尊贵,传递出豪宅的新概念。

The show flat is a three-storey villa that is few in number. Garage is located in the basement 2, and the basement is used for recreation room. The first floor is designed for living room, dining room and bedroom. In terms of layout plan, designer optimizes not only the original functions of building layout, but adds the function of display.

According to plan, designer collects the essential elements of western style and applies them to reflect the modern fashion. The designer chooses stone and wood as the main material and gives priority to the proportion of surface and body to show dignity as well as fashion in every detail. Contemporary requirements to living are revealed and satisfied by this building; obviously, the gold yellow painting, hanging crystal chandeliers, well-decorated wall and the elegant pattern on the floor can prove it. Every perfect detail delivers the new concept to define the luxury home.

以奢为美 新古典豪宅设计

ZEALOUS AND GORGEOUS THAI STYLE

热情而斑斓的泰式风情

项目名称：中颐永和（A401单元联排别墅）
设计单位：广州市韦格斯杨设计有限公司
项目面积： 330 m²
主要材料：木饰面、墙纸、布艺、石材、钢化玻璃

Project Name: Zhongyi Yonghe (A401 Unit Duplexes Villa)
Design Company: GrandGhostCanyon Designers Associates Ltd.
Project Area: 330 m²
Main Materials: wood finishes, wallpaper, fabrics, stone material, toughened glass

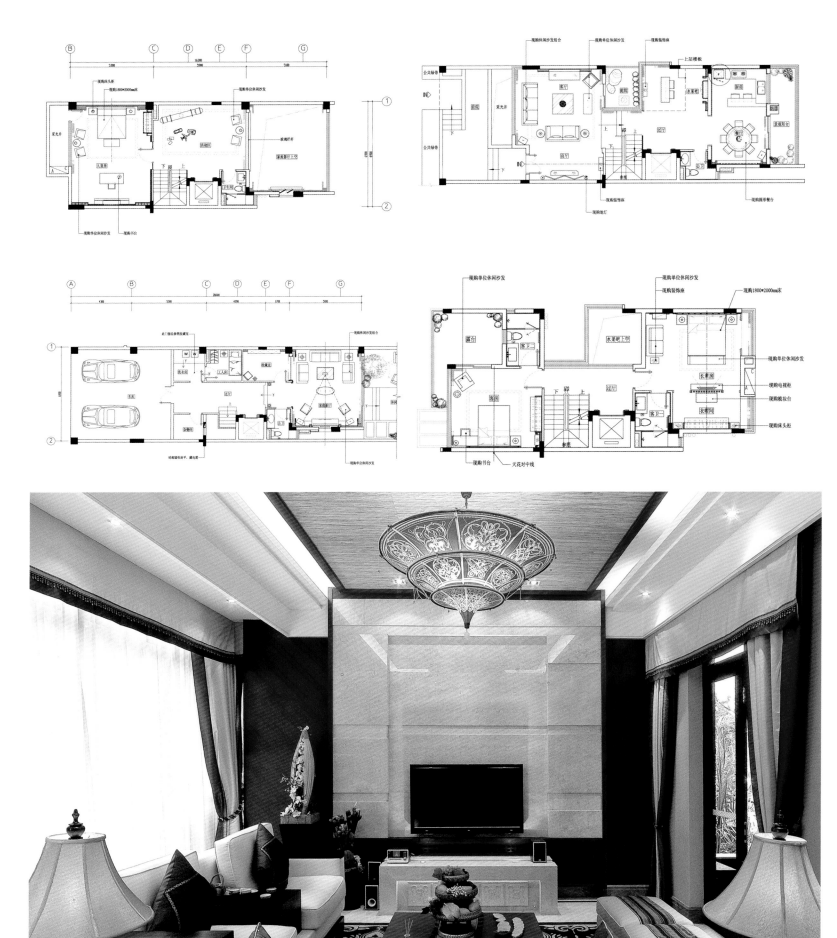

以奢为美 新古典豪宅设计

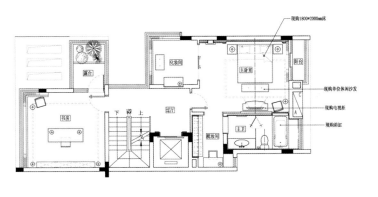

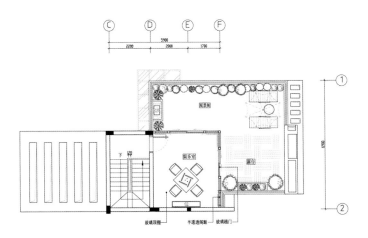

A401单元联排别墅为一个五层结构的建筑，以泰式的设计风格为主，在空间、材质、色调等方面演绎着悠闲而富有品位的生活，引领我们走进一个热情而斑斓的泰式之家。

浅色调的墙纸结合棕色的木材，与布艺家具相互搭配，用高纯度的具有鲜明地域特色的饰品加以点缀，来表现独特的泰式风情。石材与钢化玻璃在局部空间的使用，丰富了空间的层次，使材质肌理的对比鲜明，使居住其间的人们享受到一种张扬的个性。

该设计的精妙之处是木屏风的软性间隔，使整体格局紧凑且虚实相宜，让你仿佛置身于东南亚那悠闲的度假气氛之中。漫步该空间中，你会感受到在那不经意间流露出的热情与绚丽，轻松与愉悦将是你最大的体会。

A401 unit duplexes villa is a Thai-styled five-storey building which shows us a leisure living with high taste and leads us into a warm and multicolored home.

With light-colored wallpaper and brown wood going well with fabrics furniture, the unique Thai style gets expressed well decorating with the ornaments of distinctly local characteristic. The use of stone material and toughened glass in part of the space enriches the space dimension and makes the material texture distinctive, so people living in it can feel its extrovert character.

The essence of this design lies in the soft interval of wood screen, which makes the whole pattern compact and true and false fitting each other well, as if you were in the leisurely holiday's atmosphere of Southeast Asia. Walking in it, you can not only get the feeling of zeal and gorgeousness, but relaxation and delight.

THE SPACIAL SPIRIT WHEN THE MOON IN THE SKY

月亮之于夜空的空间精神

项目名称：珠海中信红树湾28-9别墅示范单位
设计单位：深圳市邱春瑞设计师事务所
主设计师：邱春瑞
参与设计：李赢
项目地点：珠海
项目面积：320 m²
主要材料：意大利木纹、实木地板、马赛克

Project Name: Zhuhai Zhongxin Hongshuwan 28-9 showflat
Design Company: Shenzhen Qiu Chunrui Design House
Designer: Qiu Chunyun
Design Assistant: Li Ying
Project Location: Zhuhai
Project Area: 320 m²
Main Materials: Italian wood grained paper, solid flooring, mosaic

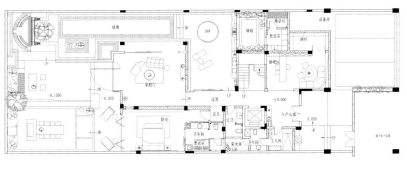
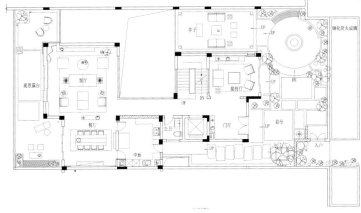
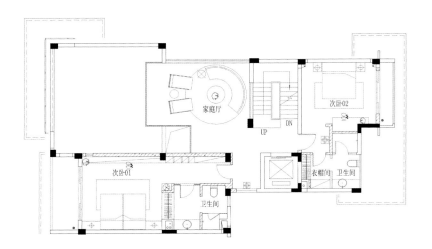
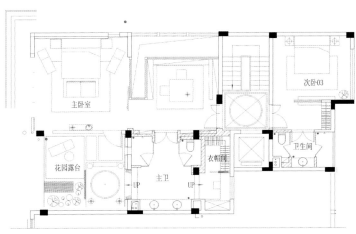

不同的空间带来的是不同的生活品质和生活氛围，人们希望藉由空间这种外在的物质形态给生活带来看不见的、内在的精神。空间的精神如同月亮之于夜空。因有精神，而凸显出空间的美好。

设计师借助多种不同的元素来打造空间的每个视觉点和面，在不同的视觉结构中，力求凸显出空间的完整性和饱满感。客厅中将光线进行了很好的"收藏"，凸显出空间的明暗对比感，增强了空间的张力。意大利木纹带来的不仅仅是清晰和时尚，一同萦绕在空间中的还有业主对生活的美好向往。

客厅和餐厅相连，在造型以及家具样式的选择上，凸显出空间的现代包容感。餐厅拥有良好的位置，旁边的窗户优化了就餐环境，也给整个空间带来了几分大自然的气息。客厅旨在结合多种文化元素，中式风范的屏风如同一位安静的淑女，站立在窗前，窗外的繁杂喧嚣，似乎在那一抹简单的水墨中顿时化为乌有。茶色的墙面和窗帘，加强了空间的宁静感受。客厅的高度决定了空间的大气与厚重。空间上下两层注重连接性和互动性，玻璃的运用，既增强了空间的通透性和现代感，同时也使空间连为一体。

二楼家庭聚会厅注重对休闲生活的营造与渲染，沙发围绕成一个半圆，是家人交流的场所。大面积的窗户提升了该区域的自然品质。简单而不张扬的吊顶设计，力求还原出一个更加真实的家庭生活空间。

书房被设置在三楼，为业主读书、学习、工作提供了一个静谧的环境，这也是书房最重要的品质。设计师在书房中依旧采用了沉稳的色调，整个空间也被合理地简化，从而让空间更加温馨。

Different space that is a kind of external matter creates different living quality and living atmosphere, with which people hope to vitalize life with invisible and internal spirit. The essence of the space resembles the moon to the night sky. With spirit, it highlights the wonderful feelings of space.

The designer connects a variety of elements to create every visual point and surface well in order to make the integrity and the feeling of fullness in different visual structure. The light in the parlor is "collected" so well to highlight the contrast of light and shade as if the space is expanded. The Italian wood grained paper brings not only definition and fashion but also the owner's yearn to the best life.

The parlor is connected with kitchen, which highlights the inclusiveness of modern from the feature and the choice of furniture style. With a good location and widows making a wonderful atmosphere of dining, the kitchen gives you a feeling of walking into nature. Parlor aims at blending multicultural elements. Chinese-style screen seems like a girl who stands in front of the widow. The hubbub of the widow outside becomes naught in the simple painting. The dark brown curtain and wall enhance the tranquility of the space. The height of parlor decides the feeling of taste and thickness. The connectivity and interactivity of up-space and down-space and the use of glass add transparent and modern sense, make the space blend into one body.

The living room on the second floor pays attention to the building and rendering of relaxation of life. Sofa is put into a semi-circle, where family members chat with each other. A large area of window enhances the natural quality. Simple and low-key ceiling aims at restoring a more real space of family life.

The study is located on the third floor, where provides a tranquil environment. This is also the most important quality of study. The designer adopts the staid tone and makes the whole space simplified rationally in order to make the space more comfortable.

ZEN RHYME. JIANGNAN STYLE

禅韵·江南风采

项目名称：福州某单层住宅
设计公司：福州华悦空间艺术设计机构
设 计 师：胡建国
项目地址：福州
项目面积：300 m²
主要材料：仿古砖、大理石、壁纸、玻璃、实木花格

Project Name: A Single Residential Apartment in Fuzhou
Design Company: Fuzhou Hua Yue art and Design Institutions
Designer: Hu Jianguo
Project Location: Fuzhou
Project Area: 300 m²
Main Materials: antique rick, marble wallpaper, glass, wood lattice

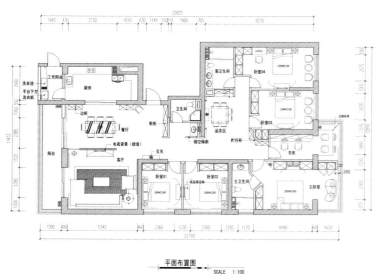

平面布置图
SCALE 1:100

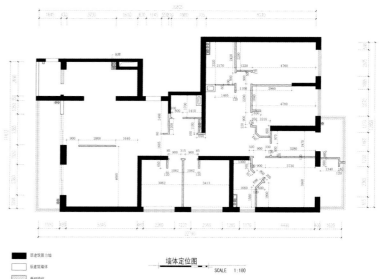

墙体定位图
SCALE 1:100

地面材料图
SCALE 1:100

吊顶布置图
SCALE 1:100

以奢为美 新古典豪宅设计

该案为面积约300平方米的单层住宅空间，整体空间较为宽敞且采光条件良好，因此，设计师围绕建筑自身具有的优势，对其展开以新东方简约主义为主题的时尚设计。

在设计过程中，设计师充分分析了使用功能与视觉效果之间的平衡关系，并努力做到古为今用。将中式传统古典建筑元素的精华所在与现代生活的时尚气息融合在一起，并将空间造型的比例、尺度、材料质感与色彩搭配和谐地组织在一起，力求能在整个设计中达到合理创新，同时能体现时代精神与人文品质的标准。

This case is designed for a single residential apartment with the construction area about 300 m². The overall apartment is spacious and with good lighting conditions, so based on these advantages, the designer adopted a new minimalist theme of oriental fashion design.

In the design process, the designer fully analyzed the balance between visual effects and functions, and strived to make the classic elements work. Combined the essence of the Chinese traditional elements of classical architecture with fashion of modern life, the designer also brought the proportion of space, scale, materials, texture and color together to achieve innovation in the rational design and reflect the spirit of the times and cultural quality standards.

DREAMS OF MING AND QING DYNASTIES

梦回明清

项目名称：南山清水溪别墅
设计单位：重庆品辰室内设计有限公司
设 计 师：庞飞、袁毅、杨万煜
项目地点：重庆
主要材料：青石、热带雨林棕石材、柚木、瓦片
摄 影 师：庞飞

Project Name: Nanshan Qingshui Villa
Design Company: Chongqing Pinchen Interior Design Co., Ltd.
Designer: Pang Fei, Yuan Yi, Yang Wanyu
Project Location: Chongqing
Main Materials: bluestone, tropical rain forest brown stone, teak, tile
Photographer: Pang Fei

以奢为美 新古典豪宅设计

251

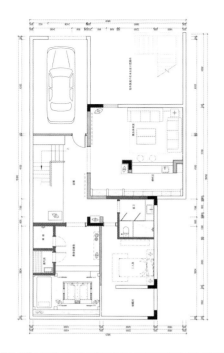
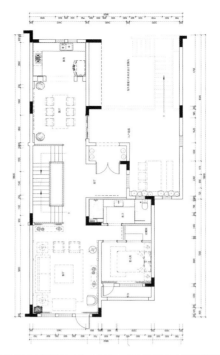
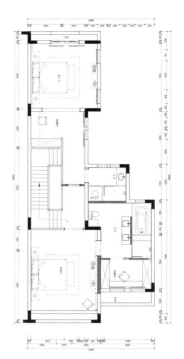

"居庙堂之高，则忧其民，处江湖之远，则忧其君。"
庄重、古典、雅致的空间使人心境宁静而愉悦，由设计师精心挑选、细心摆放的每件装饰品无不体现着大气而宽厚的心态。折扇、雀替、雕梁等传统建筑文化符号非常融洽地在各个位置相互呼应。同时，设计师又非常巧妙地运用灯光、软装饰品把古代文明与现代思潮有机地结合在一起。

在这里，"奢侈"仅仅是一个空洞的名词。建筑外墙的青砖、原木的梁柱与地面的青石，在有意与无意间填补了凡尘俗世中浮躁的心域。大量留白的墙面犹如中国传统水墨丹青中的"布局留白"，给人以无限遐想的意境，仿佛梦回到明清时代。此刻山中之清风悠然拂面，眼前浮现出"无丝竹之乱耳，无案牍之劳形"的悠远意境。身在此地，依卧罗汉床，闭目、呼吸，惬意的感觉经由身体畅达心灵。

"Ranks high temples, are concerned about the people, though stayed far away, are concerned about his lord."
Solemn, classical, elegant and tranquil space makes people pleasant. Carefully selected by the designer, each ornament placed reflects the generous and humble attitude. Folding fan, sparrow brace, carved beams and other traditional architectural cultural symbols are harmoniously laid in to echo each other very well. Meanwhile, the designer uses the lighting and soft furnishings wisely to combine the modern ideas to ancient civilizations together organically.

The word "luxury" is not just an empty term here. The bricks on the exterior walls, wood beams and bluestone on the ground intentionally and unintentionally full filled the irritable hearts in the mundane word. The large blank walls are just like the margin in traditional Chinese ink painting, which left people unlimited imaginations, as if a dream back to the Ming and Qing era. Leisurely at the moment as the mountain breeze was blowing, seems like the mood described as "no vulgar music to distract, and no paper work to fatigue". Stay here lying on bed with eyes closed, breathing, you can feel the comforts accessible through your soul.

THE VILLA COMBINED THE EAST AND WEST TOGETHER

中西精粹于一墅

项目名称：古北佘山国际别墅
设 计 师：崔龙
项目地点：上海
项目面积：600 m²

Project Name: Gubei Shenshan International Villa
Designer: Cui Long
Project Location: Shanghai
Project Area: 600 m²

以奢为美 | 新古典豪宅设计

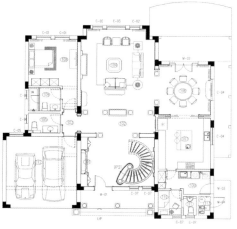
一层平面布置图 1:75

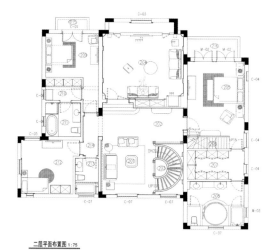
二层平面布置图 1:75

上海是东、西方文化重叠交错最彻底的城市,这种奇特的背景是新文化诞生的最大契机。早在20世纪初期,沙逊别墅、哈同花园这类"多种艺术风格混合,结构神奇的建筑"就已经见证了海派文化的兼容并蓄。

该案汲取了文艺复兴时期新古典主义风格的特质,与中华文明几百年甚至上千年的建筑精髓相扶相依,使这个现代的室内空间与所有的装饰呈现出了最自然的化学反应。这里有伊克蒂诺斯(巴特农神庙)意蕴的石柱,有曲线柔和而富于韵致的拱廊,在水晶吊灯与拱廊的辉映之下,一种独属于家的温暖弥散开来。

窗外,临佘山林语,品碧水长天;窗内,寓无限尊崇于一室,集中西精粹于一墅。这一海上独有的文化气度,巧妙绝伦地融入了这一墙、一柱、一廊,甚至一根线条之中,细细品味,仿佛每一个角落都有一个值得追溯的故事。

室小乾坤大,方寸显真情。当设计师抱着这样一种态度融入到业主的生活诉求之中,一切美好和谐图景的产生,就变成自然而然的事情了。

简约而理性的大理石拼花内蕴华美,庄重而典雅的柚木拼贴书写着品位的涵义。而主人珍藏的一石、一画、一器也仿似与这幢宅邸的装饰早早许下了约定,浑然天成。

这里的华贵之气内敛而温和,更多的是在诠释主人高尚的生活理念,有苏东坡"一蓑烟雨任平生"的豪情写意,也有倪云林艺术生活品位皆佳的名士风采。

入世,出尘,独恋这一份家的韵味。

These two kinds of cultures - Eastern and Western culture come together and immerse deeply in shanghai. It provides an advantageous opportunity to bring out the new culture. At the beginning of 20th, the Sand Som Villa and Hardoon Garden are the result of culture-integration or Shanghai Regional Culture. Moreover, these two buildings are famous for their own unique structure and affluent artistic styles.

The designer puts neoclassicism in use to this project, accompanied by Chinese thousands of years' culture. The modern room and all the decorations match each other perfectly. Under the crystal light, Parthenon pillars and graceful arcade gives you the great feelings at home.

Looking out of the window, the tender wind comes through the woods on the Sheshan hill and the clear stream flows in tranquil to the far. When turning back to the room, the application of Western and Chinese culture provides you elegance and dignity. You can feel the unique Shanghai Regional Culture from every corner of the room.

Designer makes endeavors to make the room perfect to satisfy every proprietor's requirements.

Contracted marble pattern shows the magnificence when elegant teak pattern reflects the high statues. Dulcetly, the owner's collections seem to have had romantic promise with this room tacitly.

The whole style of this project reflects an understated gorgeousness of the owner. To give you a poetical life like that Su Dongpo or Ni Yunlin is the goal that we are pursuing.

You will miss your home wherever you are.

INK PAINTING SPACE

墨情空间

项目名称：福州某住宅
设计单位：福州创意未来装饰设计有限公司
设 计 师：郑杨辉
项目地点：福州
项目面积：178 m²
主要材料：新中源陶瓷玻化砖、蒙托漆、得高软木、肌理灰砖、原木
摄 影 师：周跃东

Project Name: A Villa in Fuzhou
Design Company: Fuzhou Creative Future Decoration Co., Ltd.
Designer: Zheng Yanghui
Project Location: Fuzhou
Project Area: 178 m²
Main Materials: new source of ceramic tiles, Monto paint, De Gao cork, fabric gray bricks, logs
Photographer: Zhou Yuedong

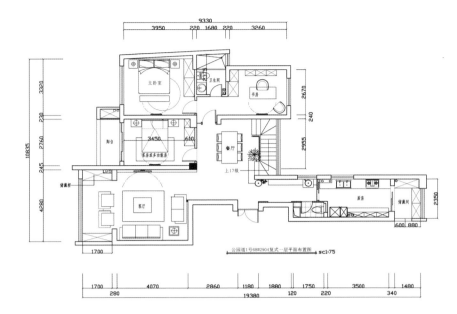

在中国土地上成长的我们,对中式建筑及室内空间的理解各有不同。在设计师的印象中,一段原木踏步、一堵灰砖墙、一组水墨画,就是设计师心中最真实的中式印象。在这个案例的设计中,设计师希望用最单纯的中式概念的抽象符号来营造新中式的空间氛围。

在设计的过程中,居住者对空间的使用要求是设计师首要考虑的一点。因此,对空间的原始布局稍微做了调整,比如将复式楼梯的位置做了改变,以使空间的结构更加合理。此外,为了使整体空间显得干净利落,将鞋柜和洗手台柜合二为一,尽力呈现视觉上的美感。

总之,该案总体上给人以和谐而纯净的印象,在这样的空间里面,能让我们深深地体会到和谐新中式的韵意。

Though grow up on China soil, we have different understanding of Chinese architecture and interior design. In the designer's mind, some still logs, a gray brick wall, a group of ink paintings forms the most authentic Chinese impression. In the design of this apartment, the designer hopes to create a new Chinese space temperament by using the most simple Chinese concept of abstract symbols.

In the design process, the requirements of the occupants is put as the priority, therefore, the original layout of the space is modified a little, such as the location change of the double staircase makes the structure more reasonable. In addition, in order to make the whole apartment look neat, the shoe cabinet and sink are combined into one, just try to make it visually pleasant.

In a word, this design presents an overall impression of harmony and purity, stay within, we could deeply appreciate the harmony of the new Chinese rhyme.

NEW YUPPIES
POLY CRYSTAL HILL

新雅皮士 POLY CRYSTAL HILL

项目名称：南京紫晶山3B 户型
设计单位：广州尚逸装饰设计有限公司
设 计 师：王赟、王小锋
项目面积：132 m²
投 资 商：保利地产

Project Name: Amethyst Mountain in Nanjing Unit 3B
Design Company: Guangzhou Shangyi Decoration Design Co., Ltd.
Designer: Wang Yun, Wang Xiaofeng
Project Area: 132 m²
Invest Company: Poly Real Estate

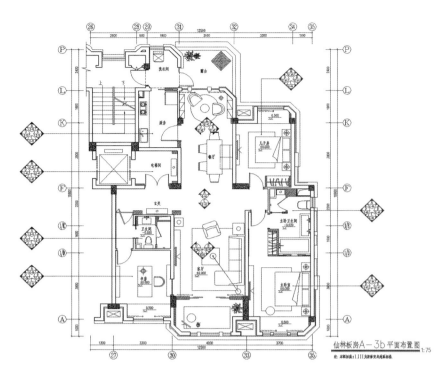

为了体现雅皮士风格的流行元素，在空间布局上，设计师采用精巧的细节设计来提升空间的层次与节奏。设计师恰到好处地把旧时代的流行元素融于现代的生活空间中，可以从空间的每一个角度体现出来。置身其中，人们可以感受到六七十年代的怀旧气息，也可以感受到现代生活的时尚品位。

设计尽量避免出现过多繁复的造型，以一些跳跃的色调作为空间的主旋律，为生活空间注入了一股怀旧的时尚气息。简易的家具与色彩到位的布艺、饰品，为空间增添了点睛的一笔。

在配饰风格上，家具与配饰的搭配非常完美，进一步体现着雅皮士风格的特点。无论是客厅、餐厅还是卧室，都摆设了白色、黑色的饰品，使简洁怀旧的家居增添了时尚、活跃的气氛。

In order to reflect the style of the popular yuppie element in spatial distribution, the design details were used to enhance the level of space and rhythm, which can be reflected from every point of space. The designer can put the pop elements from old days appropriately into modern life in space. People can feel nostalgic scenes in the sixties and seventies, and the taste of modern life style. The design avoided the excessive complicated shapes, but adopted some jumping colors as the space theme, which also embedded into some nostalgic fashion within. The simple furniture and good color fabric, accessories also become the highlights of the overall. The style of accessories matched perfectly with the furniture and accessories, which further reflects the yuppie-style features. In the living room, dining or bedroom furnishings are all in white and black, adding stylish and lively elements into nostalgic style.

BORDEAUX WINES

波尔多左岸的红酒

项目名称：波尔多左岸
设计单位：深圳市矩阵室内装饰设计有限公司
设 计 师：王冠
项目地点：深圳
项目面积：400 m²
主要材料：米黄大理石、茶镜、金箔漆、软包、墙纸

Project Name: Left Bank of Bordeaux
Design Company: Shenzhen Interior Design Ltd.
Designer: Wang Guan
Project Location: Shenzhen
Project Area: 400 m²
Main Materials: beige marble, tawny glasses, gold leaf painting, upholster, wallpaper

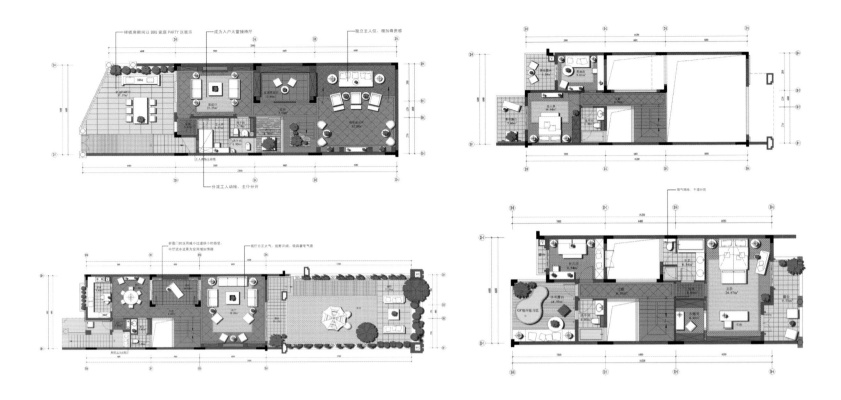

对于一栋面积达400平方米的联排别墅来说，应有的各项功能配置都应该一应俱全，设计师希望在这里展现一种别致、优雅的奢华生活。

整个空间色调丰富，设计师以驼色、杏仁色、米灰色、褐色、乳白色来映衬金色、银色及点睛的酒红色、洋红色与灰紫色。大面积的进口云石、墙纸、烫金软包面料的运用，突出了整个空间的奢华感和业主高档的生活质量。

经典的家私和配饰的运用更带来了极具亲和力的人文感受，犹如著名的波尔多岸边盛产的红酒，香醇浓郁源远流长。

This house covers 400 square meters and has complete equipments. Designer wants to give you a luxury and elegant life.

The whole room is rich in beautiful colors. Designer pays more attention to color's collocation. For example, he makes dark beige, almond, brown and ivory be matched with golden, silvery and some striking colors as wine red, carmine, and gray purple. The application of imported marble, wallpaper, gold stamping soft equipment make this room full of luxury feeling. Classic furniture and decorations bring you cultural atmosphere like famous Bordeaux Wine with its wonderful smells flowing over everyone's heart.

THE ELEGANT ANCIENT EUROPEAN STYLE

古欧洲的典雅风范

项目名称：常熟虞景山庄别墅
设计单位：巫小伟/威利斯创意设计中心
户　　型：联排别墅
项目面积：480 m²
造　　价：300万
主要材料：大理石、罗马柱、壁纸、水晶灯、欧式橱柜、木纹砖、仿古砖、马赛克

Project Name: Changshu Yu Jing Villa
Design Company: Wu Xiaowei/Willis Design Company
House Type: Townhouse
Project Area: 480 m²
Cost: 3 million
Main Materials: marble, Roman column, wallpaper, crystal chandeliers, European-style cabinets, wood tiles, antique tiles, mosaics

曾经多少次梦想回到15、16世纪的欧洲，用心去感触各种文化融合而成的古典而优雅的建筑和贵族高雅奢华的生活，而今，这一切在设计师的努力下成为现实。 入户大堂极为奢华，石材地面、穹顶、罗马柱，简单的勾勒尽显大气与典雅，奠定了大宅的基调。 入户大堂右手边为客厅，设计上干净利落，通过软装营造出高贵典雅的气氛。在颜色搭配上，既通过黑、白、灰为底色的沙发营造出优雅高贵的氛围，同时又使用华丽的窗帘体现雍容华贵之感。古铜色的壁炉更为室内添加了几分凝重感。水晶灯、花架、饰品等把这个空间装点得分外妖娆。 二层为业主的私密区，分布着主卧、次卧和书房，均以金色和银色为基调，配以或白色或深色的家具，加上吊顶、罗马柱、雕花、精致的饰品，无一不展示出欧式风格的优雅。 餐厅则较为简单，古典欧式的实木餐桌弥漫着浓郁的文化气息，皮质的座椅与餐桌相得益彰。
厨房空间较大，采用U形整体橱柜，同样洋溢着浓郁的欧式风情，雕花、金属拉手等细节体现出不俗的品质。

How many times have we dreamed back to 15 and 16-century of Europe to experience the cultural combination of classic and aristocratic elegant architecture and luxurious life. And now, thanks to the designer's efforts, all of which have become a reality. The lobby is extremely luxurious, as the stone ground, dome, and the Roman column simply outlines the elegant and ground style, which laid a basic tone of the mansion. The living room is on the right side of the lobby, and its neat design and soft decoration created an elegant style. The colors of black, white, and gray as the background of the sofa creates a style full of grace and elegance, while the gorgeous curtains adds luxurious flavor. The indoor bronze fireplace adds more dignified sense to the design. Crystal lamps, flower, jewelry, etc. decorates the space of a particularly enchanting apartment. The second floor is the owners' private area - master bedroom, guest room and study, all of which are designed based on golden and silver colors, together with the white or dark furniture and the ceiling, Roman column, carving, and fine jewelry, all of which presents European-style elegance. The dining room is relatively simple - the classic European-style solid wood dining table fully filled with cultural elements, and the leather seats and tables complement each other quite well.
The relatively larger kitchen is suitable to adapt U-style cabinet, which is also rich in European style - the carvings, metal handles and other details reflects good taste.

UNDERSTATED NEO-CLASSICISM

低调的新古典主义

项目名称：波托菲诺复式住宅
设计单位：深圳市朗昇空间艺术设计有限公司
设 计 师：袁静、钟建福
项目面积：330 m²
主要材料：地砖、地板、墙布等

Project Name: Duplex Apartment in Portofino
Design Company: Shenzhen Lonson Environmental Art Design Co.,Ltd.
Designer: Yuan Jing, Zhong Jianfu
Project Area: 330 m²
Main Materials: floor tile, floor board, wall cloth, etc.

这是一个三层复式的户型，面积有330平方米，称之为低调的新古典主义设计风格。这是因为其中蕴含有欧式、美式、东南亚、中式等元素，多种风格相互交糅在一起，旨在营造一个轻松自然、质朴无华、灵活多变的宜居环境。细细品味，每个角度都有风景，每个空间都有特点，似乎在暗示生活在其中的人们，每天都有不一样的精彩故事在上演。

首先，入户花园已被改造成为主人生活的重要空间——茶室。里面摆放有木雕、藤制沙发、实木书柜、木箱等，茶几与沙发原本是旧家私，现在旧物翻新，但是却一点也感觉不到它的陈旧，反而给整个空间增添了不少生活情趣。无疑，这里是一个静思悟道、会客闲聊、品茶遐想的极佳场所。

This three-storey duplex apartment covers 330 square meters and is designed in the style of understated neoclassicism. European style, American style, South Asian style and Chinese style permeate together into this project to create relaxed, natural, and flexible environment for living, which leading to the naming of understated neoclassicism. It seems that fresh story wonderfully happens every day in this place when enjoy the beautiful scenery from every corner and feel the unique style in every part of the apartment.

Above all, indoor garden has remade into tearoom that plays an important role in the owner's life. The tearoom is furnitured with woodcarving, cany sofa, solid wood bookshelf and wooden case. The old teapot and old sofa are renovated, surprisingly, they create more pleasure for life abandoning the feeling of behind the times. Undoubtedly, this tearoom is a really fabulous place for meditation, meeting friends, tea tasting and being lost in the fanciful thoughts.

客厅空间保持了原结构的高挑，显得大方而开阔。客厅家私实际上也是旧物翻新的，布料颜色的选择沿用与茶室同样的浅绿色系，生活气息比较浓厚。由于客厅空间较大，故又增添了部分美式家私，如靠椅、酒吧椅、吧台等，显得沉稳大方，再辅以精心搭配的软装细节，细腻而有品位，极富生活情调。

私人空间温馨舒适，儿童房以浅蓝色与白色为主色调，与拱门造型相得益彰，透出了热带海洋的生活气息，整体上显得干净、清爽，是令儿童们充满幻想的空间。

主人房面积较大，故将书房纳入其内。厚重的美式家私摆设在其中，阔绰有余。同客厅一样，房间里不乏精心搭配的软装细节，色彩温馨、性情细腻，充满着浓厚的生活情调。

The structure of living room remains the original height to look elegant as well as wide. Actually, the furniture here is also the production of renovation from old ones. When picking up the cloth, the designers choose the light green to be corresponding with color scheme of tearoom, which injects more vitality into daily life. In order to make full use of commodious living room, the designers bring in American furniture, like armchair, bar-bench, and bar counter. All of this elaborate design accompanied by soft equipment, which marks the elegant taste of the owner.

Private room should be comfortable. Blue and White is the main color, matching the arched door well, full of the smell of Tropical Ocean. Cleaning and refreshing atmosphere leaves children enough fantasies to chase.

The main bedroom is large enough to share a corner with study room. American furniture inside is shining the glories of dignity. Similar to living room, the designer adopts soft details to decorate the bedroom, with cozy colour and smooth nature going well with life.

SPIDER MAN AND JAZZ

蜘蛛侠与爵士乐

项目名称：嘉宏湾花园样板房
设计单位：深圳市凌奔环境艺术有限公司
设 计 师：凌奔
项目地点：深圳
项目面积：148 m²
主要材料：石材、墙纸、镜面不锈钢、夹丝玻璃

Project Name: Cavendish Beach Gardens Model House
Design Company: Shenzhen Ling Ben Environmental Art Co., Ltd.
Designer: Ling Ben
Project Location: Shenzhen
Project Area: 148 m²
Main Materials: stone, wallpaper, mirror stainless steel, wired glass

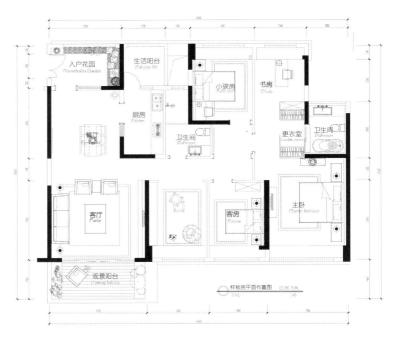

《蜘蛛侠3》里面有一段爵士乐，音乐充满激情且富有感染力，演绎出了主人公内心的情感世界，引领着新鲜而激动人心的爵士乐风格的未来走向。在主人的房间中，也挂着蜘蛛侠的海报照片，正如该居室的主人一样，充满着桀骜不驯的性格特征。

该案在材质的运用及搭配上进行了创新，使得这个空间在保持原有简约基调的基础上亦不失奢华感，打破了人们对传统现代风格的固有认识。淡金色墙纸营造出温馨的居室氛围，水银镜饰面及镜面不锈钢的运用，为空间添加了现代时尚的气息，同时使空间变得通透灵动，皮革材质的运用使奢华感升级，淡雅的饰品柔化了空间硬朗的线条。

在完美的配色以及温馨的居室氛围中，我们不禁为生活的美好而欢欣鼓舞，一曲浪漫的爵士乐是这个空间最好的注脚。

The jazz from "Spider-man 3" is passionate and appealing, which is the interpretation of the protagonist's inner emotional world, and also leads the new and exciting jazz style in the future. A Spider-man poster is hanging in the master bedroom, just like the owner of this room, who has full of rebellious character traits.

For the design of this apartment, the innovative materials are used to make the apartment maintain the original succinct style with luxury style, which breaks people's traditional knowledge on modern style. The pale golden wallpaper creates a warm room, and the use of mercury mirror and stainless steel mirror adds some modern and stylish elements within, at the same time, the overall design becomes transparent and lively. The use of leather materials upgrades the sense of luxury, and elegant decorations soften the lines of tough style.

In the living space with perfect color combination and warm atmosphere, we can not help rejoicing for the good life, and the romantic jazz is the best footnote in this apartment.

PLAY THE BEAUTIFUL ROMANTIC MUSIC

奏响浪漫的华美乐章

项目名称：深圳大东城·嘉宏湾花园
设计单位：深圳市尚邦装饰设计工程有限公司
设 计 师：潘旭强、潘冬东
项目地点：深圳
项目面积：128 m²
主要材料：大理石、涂料、金属、玻璃

Project Name: Shenzhen East Bay Garden Carven
Design Company: Shenzhen S+B Design Studio Co.,Ltd.
Designer: Pan Xuqiang, Pan Dongdong
Project Location: Shenzhen
Project Area: 128 m²
Main Materials: marble, paint, metal, glass

平面布置图
比例 1:75

在这套样板房中，设计师用最引人注目的材料来包裹这个家的全身，不论在白天还是夜晚，装饰品剔透的质感和夺目的外观为家注入了无限活力，使这个家居空间展现出无穷的魅力。

在主人房与儿童房的设计中，强烈地体现出居室主人的性格以及身份特征。主人房中，用稳重、浪漫、大气的色彩表现主人不俗的欣赏品位，男孩房中的趣味性，女孩房中的梦幻气质，无论是在色彩表现方面，还是在细节与装饰上，都使这个家的设计在欧式特有的华丽中延伸出家的乐趣。

In this model room, the designer adopts the most eye-catching material to wrap the whole house. Whether during the day or night, the texture and eye-catching appearance of the ornaments embeds infinite energy to the house day and night, and makes this living environment display endless charm.
The design of the master bedroom and children's room strongly reflect the character and identity of the master. The colors used in the master bedroom are steady and romantic, displays the master's good taste. The boy's room is fun, while the girl's room is dreamy - both in color, details and decoration, all of which make the unique home appear the magnificent European-style as well as an extension of fun.

NEW DEFINITION OF LUXURY AND ROMANCE

奢华与浪漫的新符号

项目名称：安徽华地公馆B1-1样板房
设计公司：深圳市墨客环境艺术设计有限公司
设 计 师：王勤俭
项目面积：127 m²
主要材料：白玫瑰大理石、香槟红大理石、黑金花大理石、雅士白大理石、不锈钢马赛克、回纹马赛克、车边银镜、水银镜磨花、墙纸、绒布软包、密度板锣花、哑银箔纸

Project Name: Anhui Huadi B1-1 Showflat
Design Company: Shenzhen Moke Enviroment Art Design Co., Ltd.
Designer: Wang Qinjian
Projece Area: 127 m²
Main Materials: White-rose marble, Champagne Red, Black gold flower marble, white marble, stainless steel mosaic, fret mosaic, silver mirror, wallpaper, soft equipment, high density board, foil paper

该案想传递一种低调奢华的生活方式，既有欧式生活的贵气与精致，又无金碧辉煌的浮华。整套大宅的设计理念是奢华主义与现代浪漫主义的完美结合。流线形与直线相互搭配，现代而奢华的家具精美华丽，明亮的色彩相互交错。所有豪华的家饰，个性的墙纸相互呼应，此情此景充斥着浪漫的诱惑力，表现出王者的风范。设计者试图在设定的消费群中，将一种华丽贵气的表达形成另一套符号系统，将物体还原到最浪漫、最纯粹的状态……

该案中，合理的空间格局改造使得空间的序列完整。经过调整后，门后正对玄关案台，左侧是客厅，右侧是钢琴厅，再结合磨花玻璃、磨花镜的虚拟反射，使得空间奢华而大气。

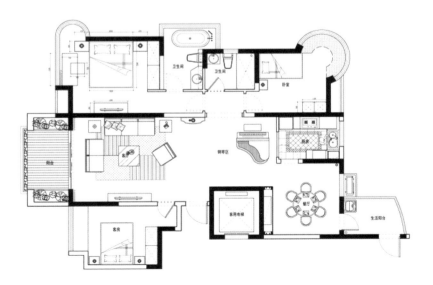

Model home aims at providing an understated-grand life, which contains European elegance but no over-decorated vanity. The purpose of the whole design is to show combination of luxury and modern Romanism. Designer turns all the magnificence into the natural and pure ones.

In this case, reasonable spatial pattern make the whole space orderly. After adjusting, the door is right in front of the porch, and living room is on the left; piano room is on the right. The reflection of the polish glass makes the room more luxurious.

以奢为美 | 新古典豪宅设计

WITH REGARD TO SPACE AND BEAUTY

关于空间 关于美

项目名称：简欧风格·珀丽湾别墅
设计单位：广州道胜装饰设计有限公司
设 计 师：何永明
项目地点：高要
项目面积：310 m²
主要材料：泰柚直纹木、大理石、陶瓷锦砖

Project Name: Succinct European Style - Poli Bay Villa
Design Company: Guangzhou Daosheng Decoration and Design Co., Ltd.
Designer: He Yongming
Project Location: Gaoyao
Project Area: 310 m²
Main Materials: teak, marble, ceramic mosaic

以奢为美 新古典豪宅设计

首层平面图

二层平面图

三层平面图

信业肇庆珀丽湾一期别墅，位于肇庆高要。现代欧式风格，不仅雍容华贵，同时又兼顾典雅浪漫与时尚现代感，反映出独特的个性，体现出主人高雅而含蓄的性格特征。

在这个空间中，设计师特意放置了一些雕塑作品，为了让这些作品能够进一步融入这个生活空间中，设计师以白色为主基调，淡化空间的线条、色彩与装饰，让雕塑本身的张力成为空间的重点。

玄关后的整个公共区域中，大面积的玻璃窗具有延伸视觉的作用，也使室外的光线充分进入，室内外的空间连为一体，形成了一种有趣的对话姿态。空间里的一物一景，都能彼此烘托，彼此唱和。而这一切，又使空间重新回到"人"本身。关于生活，关于美，才是空间里最该被在乎的一切。

Xinye Zhaoqing Poli Bay Villas are located in Gaoyao Zhaoqing. The modern European style is not only elegant, romantic, at the same time embedded with the modern fashion and reflected the unique personality, which reveals the owners' elegant and humble character.

The designer deliberately places a number of sculptures in the apartment in order to make these works further integrate into the living space. The designer uses white tone to balance the lines, colors and decoration, so the tension of the sculpture becomes the highlight.

The large window with visual extension effect is set at the public entrance area, which also makes it access to the outdoor light, so the indoor and outdoor are brought together and become an interesting dialogue. Everything in the apartment is a contrast with each other and echoes to each other. All make the center back to "people" itself. Life and beauty should be cared the most among everything.

以奢为美 新古典豪宅设计

MODERN TIMES

摩登时代

项目名称：福州某摩登住宅
设计单位：福建国广一叶建筑装饰设计工程有限公司
设 计 师：李超
方案审定：叶斌
项目地点：福州
项目面积：135 m²
主要材料：紫罗红大理石、镜面玻璃、艺术马赛克、布纹硬包、钨钢

Project Name: A Modern Residential Apartment
Design Company: Fujian Guoguang Yiye Building Decoration Design Engineering Co., Ltd.
Designer: Li Chao
Program Validation: Ye Bin
Project Location: Fuzhou
Project Area: 135 m²
Main Materials: Rosso Lepanto marble, mirror glass, art mosaic, cloth pattern hard pack, tungsten steel

进入室内，最先映入眼帘的是餐厅区，其区域为半弧形，是一种低调奢华的风格。圆形的茶色吊顶上一盏流苏状的黑色水晶灯散发出深浅不一的光泽。与餐桌相对的，是一个储物柜，咖啡色的色调沉稳且富有质感。

从餐厅往右，便来到了客厅。客厅户型方正，墙面的设计是一个亮点。电视背景处，设计师将大理石和茶色镜面进行搭配，对二者进行了相同肌理的处理。与电视墙遥相呼应的沙发墙，以咖啡色和米色的软包进行装饰，以条纹增加了空间的视觉长度。客厅的家具选择了很有个性的黑色皮质沙发，犹如黑白电影一般，具有浓浓的复古韵味。

餐厅的左侧是卧室区。深色的实木地板迎合了摩登时代的华丽气质。卧室被一堵墙分隔为更衣间和睡眠区。卧室入门的一侧墙面延续了餐厅的墙面装饰风格，旁边摆上一盏时尚的落地灯，搭配上皮质单人沙发，稳重之中又蕴含一些俏皮。在灯光的映射下，即使是黑、白、咖啡色这些简单的色彩，也能变幻出丰富的层次。

The first thing can be seen of this design is the dining area, which is semi-arc and with low-key luxury style. The fringed black crystal lamp on the round brown ceiling emits different shades of gloss. Opposite to the table is a locker in brown color with quite sense of texture.

Right to the dining room is the living room. Living room is built in square, the design of the wall is the highlight. At the background of the television, the designer matches the marble and tinted mirror by putting the two in the same texture processing. Echoing to the TV background wall, the sofa background wall soft-decorated in brown and beige is to enhance the length of the visual space. The furniture selected in living room is the personalized black leather sofa, which seems like a black and white film with a deep sense of retro charm.

Left to the dining room is the bedroom. Dark wood floors meet the modern era of gorgeous temperament. The bedroom is separated by a wall for the locker room and sleeping area. The wall at the entrance of the bedroom continues the restaurant's wall decor style - a stylish floor lamp is set beside a single person leather sofa, which embeds cuteness into ground style. In light, even the basic colors like black, white, brown can also have rich changes.

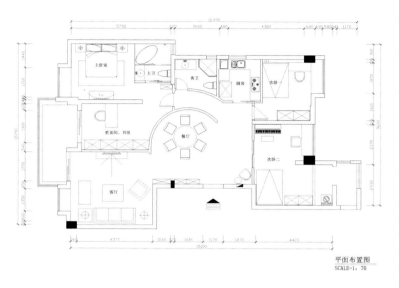

平面布置图
SCALE=1：70

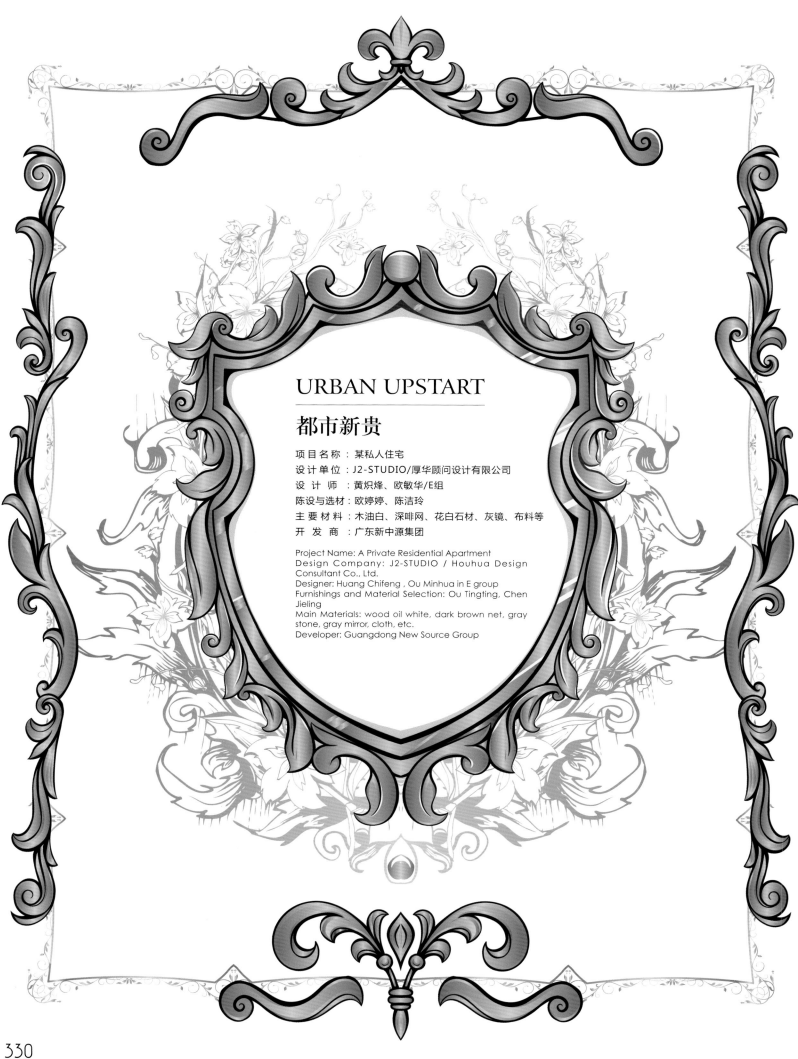

URBAN UPSTART

都市新贵

项目名称：某私人住宅
设计单位：J2-STUDIO/厚华顾问设计有限公司
设 计 师：黄炽烽、欧敏华/E组
陈设与选材：欧婷婷、陈洁玲
主要材料：木油白、深啡网、花白石材、灰镜、布料等
开 发 商：广东新中源集团

Project Name: A Private Residential Apartment
Design Company: J2-STUDIO / Houhua Design Consultant Co., Ltd.
Designer: Huang Chifeng, Ou Minhua in E group
Furnishings and Material Selection: Ou Tingting, Chen Jieling
Main Materials: wood oil white, dark brown net, gray stone, gray mirror, cloth, etc.
Developer: Guangdong New Source Group

该案中，整体空间以白色为基调，渲染出一个纯净的古典空间氛围。在材质的选择上，设计师大量运用白色饰面和银灰色的布料作为墙身的主材，以深色的石材和灰镜造型点缀出极具观赏性的艺术造型。艺术品的运用也是空间装饰的重要元素之一。

在整体空间的布局上，设计师以大空间的划分为主，将原来四个卧室中的其中一个变成了独立的书房，书房与餐厅之间运用了通透的屏风作为分隔，使整个户型的空间关系变得紧凑、大气。各个空间都有独特的艺术元素。

The overall design is based on the white tone, which renders pure classical atmosphere. For the choice of materials, the designer makes extensive use of white and silver finishes cloth to the main wall, decorated with dark gray stone and mirror to shape of a highly ornamental art form. The use of decorative art is also one of the most important elements of the design.

For the overall layout, the design is mainly based on the division of a large space - one of the four original bedrooms is separated as an independent study, and a partition is used between the study and the dining room, so that the whole apartment becomes more compact and open. Each room has its own unique artistic elements.

图书在版编目（CIP）数据

以奢为美：新古典豪宅设计 / 陶陶主编. -- 南京
：江苏人民出版社，2012.7
　ISBN 978-7-214-08017-2

Ⅰ. ①以… Ⅱ. ①陶… Ⅲ. ①别墅－室内装饰设计－
中国－图集 Ⅳ. ①TU241.1-64

中国版本图书馆CIP数据核字(2012)第039180号

以奢为美——新古典豪宅设计

中华文化 策划　陶陶 主编

责任编辑：	陈丽新　蒋卫国
装帧设计：	张　萌
排版设计：	李　迎
责任印制：	彭李君
出版发行：	凤凰出版传媒集团
	凤凰出版传媒股份有限公司
	江苏人民出版社
	天津凤凰空间文化传媒有限公司
销售电话：	022-87893668
网　　址：	http://www.ifengspace.cn
集团地址：	凤凰出版传媒集团（南京湖南路1号A楼　邮编：210009）
经　　销：	全国新华书店
印　　刷：	利丰雅高印刷（深圳）有限公司
开　　本：	965 mm × 1270 mm　1/16
印　　张：	21
字　　数：	168千字
版　　次：	2012年7月第1版
印　　次：	2012年7月第1次印刷
书　　号：	ISBN 978-7-214-08017-2
定　　价：	318.00元（USD59.00）

（本书若有印装质量问题，请向发行公司调换）